Mohamed Abdel-Raheem
Eva A Ajaj
Sajida Saeed Aziz

Artificial Intelligence in Science

Mohamed Abdel-Raheem
Eva A Ajaj
Sajida Saeed Aziz

Artificial Intelligence in Science

Noor Publishing

Imprint

Cover image: www.ingimage.com

Publisher:
Noor Publishing
is a trademark of
Dodo Books Indian Ocean Ltd. and OmniScriptum S.R.L publishing group

120 High Road, East Finchley, London, N2 9ED, United Kingdom
Str. Armeneasca 28/1, office 1, Chisinau MD-2012, Republic of Moldova, Europe
Managing Directors: Ieva Konstantinova, Victoria Ursu
info@omniscriptum.com

Printed at: see last page
ISBN: 978-620-9-57538-9

Artificial Intelligence in Science

By

Prof. Dr. Mohamed Abdel-Raheem

Assist. Prof. Eva A Ajaj

Lecture. Sajida Saeed Aziz

(2026)

Artificial Intelligence in Science

Prof. Dr. Mohamed Abdel-Raheem Ali Abdel-Raheem

Professor of Entomology (Biological Control)

Pests & Plant Protection Department

Agricultural and Biological Research Institute

National Research Centre.

Assist. Prof. Eva A Ajaj

Internal and Preventive Medicine, Clinical Pathology, College of Veterinary Medicine, University of Mosul, Iraq

Lecture. Sajida Saeed Aziz

Department of Horticulture, College of Agricultural Engineering Sciences, University of Sulaimani, Iraq

Lecturer Fuad M. Ahmad

Department of Soil and water, College of Agriculture Engineering Sciences, Salahaddin University-Erbil, Iraq.

Dr. S.R. Rajkumar

The Head & Librarian at the Rev. Fr. S.J. Thomas EJ Library & Information Centre, Pilathara, affiliated with Kannur University, Kerala, India

Lecture. Viyan Ghazi Mohammed

Plant protection Department, College of Agricultural Engineering Science, Duhok University, Kurdistan, Iraq

Content list

Introduction

Artificial Intelligence, or AI, is the field of computer science dedicated to creating systems capable of performing tasks that would normally require human intelligence. It involves creating algorithms that enable systems to analyze and interpret vast amounts of data. These algorithms then allow computer systems to imitate aspects of human thought, enabling machines to replicate tasks such as learning, planning, problem-solving, and decision-making.

With the changing climate and strained resources, whether that be financial or labor shortages, the water industry is facing significant challenges. Extreme weather events are making water increasingly scarce, unpredictable, and polluted, leading to more stress on our water infrastructure, treatment processes, and those who manage it all.

Chapter (1): Use of Artificial Intelligence in management of insect pests

Mohamed Abdel-Raheem[1] and S.R. Rajkumar[2]

[1]Pests & Plant Protection Department, Agricultural and Biological Research Institute, National Research Centre, 33rd El Buhouth St, (Postal code: 12622) Dokki, Giza, Egypt.

E-mail: abdelraheem_nrc@yahoo.com, abdelraheem_nrc@hotmail.com

Tel: (+2)01155527583, (+2)01009580797

[2] Head & Librarian,Rev.Fr.E.J.Thomas .SJ.Library & Information Centre,St.Joseph's College,Pilathara,Kannur University ,Kerala ,India

Abstract:

Insects lead to losses of 30 to 40 percent of crop products annually all over the world. To ensure that there is no excessive loss, it is necessary to identify, classify, and control such pests. These processes are time-consuming and ineffective to implement in a manual way. The conventional methods are often not sufficient to deal with these dynamic behaviours of pests, which damage crops and result in more chemical use. As a result, the use of Artificial Intelligence (AI) methods in identifying and controlling pests provides a possible way out, as a solution to the challenges of changing pest populations and the need to find sustainable solutions in agriculture. AI offers a radically new method of utilization of sophisticated algorithms to process diverse data patterns of various rates based on numerous sources, including sensors and imagery. This will allow efficient pest identification, early detection and predictive modelling, leading to improved depiction of decisions to be made in managing pests by reducing unselective use of pesticides and optimization of interventions. AI can reduce costs of economic losses, but also advance non-polluting methods of managing pests in a more efficient and resilient system.

Key words: Artificial Intelligence, Machine and deep learning, Sensors, Pest detection, classification, Monitoring

---

Introduction

It is well-known that insect pests are a significant cause of agricultural losses throughout the world. Agricultural losses can be minimized with the help of prevention and control of these pests. Precursors of successful work in this area are the competent identification and categorization of insect pests, and the end goal is to identify the species, identify its host plants, and assess the pest control measures required. Automatic insect pest recognition has recently become a topic of frenzy, due to the fact that this method demands constant and expensive surveillance. The conventional monitoring is based on the services of skilled professionals observing the pests by hand; this approach is more error-prone, labour-demanding and restricts the scope of other applications [1-5]. Moreover, the carefree use of pesticides often leads to unwanted harm to the valuable organisms and the ecosystems. Considering such challenges, artificial intelligence is a revolutionary change in that it is able to create interventions that are specifically aimed at limiting the impact of ecological processes and improving effectiveness. The use of AI, which can be combined with cameras and the internet, creates a novel solution of automatic pest recognition and tracking in the contemporary agricultural setting, thus contributing to the enhancement of the efficiency of insect surveillance devices in agriculture significantly (Fig. 1) [67].

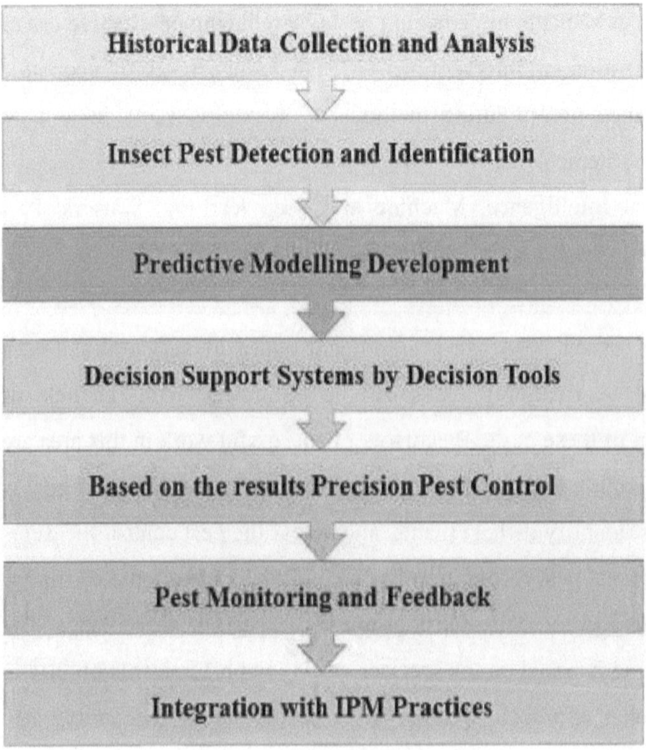

Fig. (1): Working model of Artificial intelligence in pest detection and further use as a tool in integrated pest management (IPM).

The successful use of the artificial intelligence technology in the sphere of pest management in agriculture and forestry has become a promising field, which provides new solutions that are more efficient, accurate, and eco-friendly. Due to the rapid advancement of AI in science, theories and technologies associated with AI like Smart Pest Monitoring (SPM) are a relatively new field of scientific knowledge in the area of Integrated Pest Management (IPM). The SPM is supposed to improve the automated and intelligent data gathering related to the major crop insect pests, and as a result, increase the possibility to monitor the pest and provide early warnings by means of the combination of AI, the Internet of Things (IoT), big data, and other modern information technologies and tools. The identification and identification of insects, as well as the theories associated with their density analysis, can help in making decisions on the management processes like pesticide spraying and other management processes that may help to minimize losses and increase food production [8–10]. AI is a broad concept and it includes such strategies as machine learning, computer vision, and data analytics, which are used to analyze massive and difficult datasets. This has been capable of offering an unprecedented upper hand in addressing pest related problems as it improves precision and prediction precision. The combination of AI and pest management promise to eliminate the need to use chemical interventions, decrease the environmentally harmful effect, and maximize the use of resources. Although the future of AI in pest management has a positive perspective, some issues remain, such as the issue of data privacy, legal adherence, and the need to work across disciplines, such as agronomy, ecology, and computer science. Moreover, it is essential to make the AI technologies accessible to small-scale farmers and areas with a shortage of resources to ensure that the advantages have been enjoyed by many people regardless of the type of agriculture in different areas [11,34]. In brief, there is a significant possibility of revolution in the field of insect pests management in agriculture, horticulture, and forestry with the introduction of AI. By leveraging AI-powered insights and technologies, the stakeholders will be able to adeptly address the risks of pests and improve productivity, as well as to ensure the environmental sustainability of food and fibre production systems. With the use of AI, the pest management system can move to the proactive, smart approach to the reactive one, keeping the food systems around the world safe and reducing the environmental footprint to a minimum. In this review, the complex value of AI in pest control is discussed, and insights into why the AI-driven system is useful in the detection of pests, effective decision-making and sustainable agricultural practices, as well as benefits and limitations of AI, are presented through the analysis of technological developments [15, 16]. Categorization of artificial intelligence. The categorization of AI is carried out using an array of methods, that is, functionality, learning method, and sphere of application. To begin with, according to functionality, there are two major branches of AI: narrow AI (also known as weak AI) which can perform certain tasks or a set of tasks; it is extremely efficient in its specialized field, but it cannot be considered intelligent. This can be in the image-recognition systems, virtual assistants and recommendation algorithms. Secondly, general AI (also known as strong AI) demonstrates intellectual capabilities similar to those of humans and can be used in a variety of areas by processing and learning. It is capable of changing to suit most tasks and scenarios and possesses the intelligence similar to human beings [17–18]. Based on the learning strategy, AI falls under three broad groups, supervised learning, where models are trained with labelled data, which includes input and output, the system learns to make decisions or predictions by learning input-output pairs. Secondly, there is unsupervised learning (also known as unlabelled learning) which is used to learn the patterns, structures or associations in the data without being taught what they are. Third, the reinforcement learning, that enables an agent to learn the best decision making strategies by the means of interaction with an environment, where the agent receives feedback, in the form of penalties or rewards, that helps direct the agent towards maximizing cumulative rewards in the long term [19-21]. Indicatively, to classify 22 insect pests, Kar et al. [22] provide a self-supervised learning (SSL). The experiment compared the performance of segmented and raw in-field images on the application of differentSSL methods, such as Barlow Twins, Bootstrap Your Own Latent, and Nearest-Neighbour Contrastive Learning of Visual Representations (NNCLR). All the SSL methods were used with the ResNet-18 and ResNet-50 models through the application of feature-specific image training on original RGB images and foreground segmented images, which had been pre-trained on original RGB images [22]. Besides, according to the area of application, AI is categorized into three types, namely, Natural Language Processing (NLP), which allows computers to understand, read, and write human language; it has such applications as sentiment analysis, language translation, and chatbots. Another category is computer vision which supports analytical and interpretive visual data processing of an image or video, and can be used in facial recognition, medical image analysis, object detection, and autonomous vehicles. The third category is robotics, which includes the design and development of robots that can do tasks independently or in collaboration with humans; they are used in such areas as industrial automation, as well as in healthcare support and space exploration. A typical example of a web-based tool that can be utilized under this classification can be referred to as a tool that can use NLP technology to help managers and farm workers in detecting diseases or pests in their crops. The program suggests the probable cause of the given problem that is represented in the form of a sorted list, and the recommendations of control measures, in case of the end users provide a text description of the observed symptoms in the plant. The classification system gives an idea of the different functionalities, learning methods, and areas of application of AI [22–26]. This is achieved by using experience as one of the main learning and enhancement platforms. Deep learning (DL) is a more specific form of ML and part of AI, and it entails multilayered neural networks, also known as deep neural networks. DL has been designed to replicate the information processing system of the human brain or animal brain and thus it allows the learning and decision-making process with minimal human intervention [27-29]. ML in the context of pest detection and monitoring provides powerful tools in the analysis and interpretation of different forms of data such as images, sensor data and environmental data in order to detect and control pest infestation. The next section contains the description of the way ML could be useful in pest detection and monitoring and those tools that have been designed to this end [30]. Image and classification recognition. The ML algorithms can be trained to identify and label the images of crops and detect plants that are damaged by pests or even pests themselves. Such devices as drones with high-resolution cameras can be used to take pictures of farmlands. An example is the mobile application known as the Plant Village invented by Penn State University which uses the ML to detect pests and plant diseases according to the images uploaded by the user. Similarly, the ML and shape features were used to classify 24 insect classes with the use of the Wang and Xie dataset [31-34].

IoT machines and sensor networks

The data provided by the sensors in agriculture fields can be analysed by ML algorithms to identify a pattern related to the activity of pests (Fig. 2). The devices like the IoT equipped with sensors to measure soil moisture, humidity, temperature, and other environmental parameters can provide real-time data to be analyzed by the ML. The AI algorithms can then be used to track the population of pests in real time and give early warning. Spornado insect monitoring system is a system that makes use of the ML algorithms to estimate and detect flying insects through sensors placed on insect traps to help the farmers to monitor the population of pests. In a similar manner, the Trapview system, which was developed by EFOS

Ltd., utilises the IoT-enabled traps used to attract insects, which are fitted with cameras and sensors, and the data collected is then analysed by AI algorithms to track the population of pests in orchards and vineyards and hence cause timely intervention [35-38].

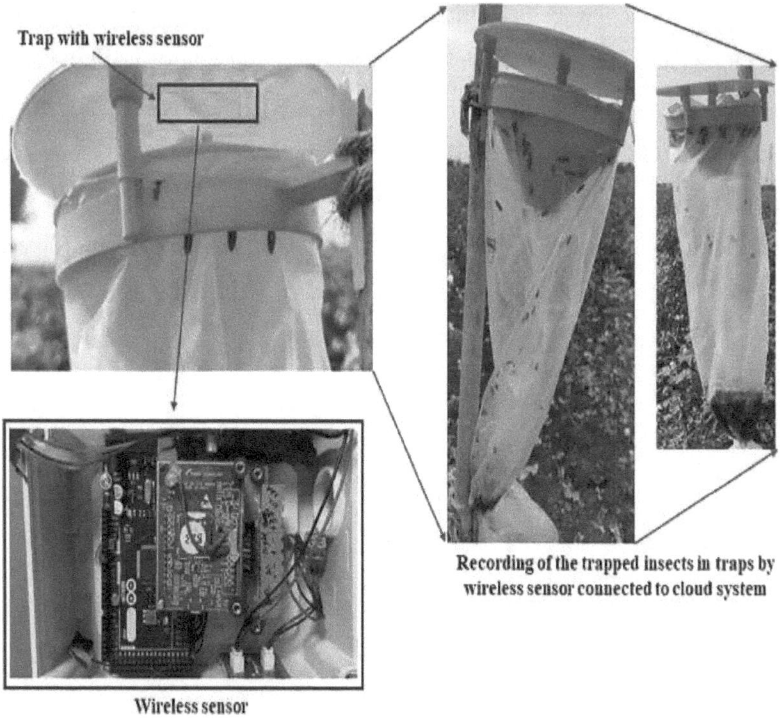

Fig. (2): Insect trap with camera mounted sensor; real time insect trapped in device are recorded by image sensor.

Machine-learning algorithms can be used to identify the pest species automatically with the help of various data sources, such as images, DNA sequences, and acoustic signals. The traps can be Internet-of-Things (IoT)-enabled and record images to identify the pests by sending the data to central systems. These automated processes are able to identify pests much faster than human processes. Improved algorithms are used to increase the level of pest identification and minimize misidentification errors. Literary devices like handheld devices or smartphone applications with ML-infused identification algorithms would be able to deliver quick and precise field identification (Fig. 3). The PestID application was created by the University of Georgia and uses machine-learning to recognize insect pests based on an image of the smartphone camera. Spiesman et al. could detect 36 species of bumble bees by using the ML tools and they could detect Lepidopteran insects with 75□ percent accuracy with the help of Angel invariant Gabor algorithm [39- 44].

Fig. (3): Identification insect pest species based on images by using AI smartphone application.

Predictive Modelling and Pest Forecasting. The machine-learning models can be trained on the chronologically ordered data related to pest outbreaks, crop type, weather conditions and other factors related to them to predict future pest outbreaks. These types of forecast models enable farmers to implement effective interventions and minimize the destruction of crops by applying the best practices, e.g., timely use of insecticides or planting of resistant crops to the pests. Markovic et al. established the effectiveness of machine-learning algorithms to predict the population dynamics of insect pests in agricultural landscapes and reached accurate predictions of pest outbreaks with the use of various datasets. The Commonwealth Scientific and Industrial Research Organisation (CSIRO) of Australia created a machine-learning system named Pest Forecast, which is used to predict pest behavior using meteorological predictions and previous data, thus allowing farmers to take pre-emptive action [45- 48]. Deep Learning (DL) Deep learning involves the use of the multilayer neural networks to derive higher-level features of the raw input data. Deep learning has already shown a lot of success in various areas of application, such as the field of natural language processing, object recognition, and the detection and monitoring of pests in the agricultural and urban settings [49]. Image Analysis by DL Models Convolutional Neural Networks (CNNs) are the common examples of DL models that are effective at visual analysis and identifying patterns in images. The CNNs are exclusively tailored to visual data processing and analysis video and photographs. To classify pest images, CNNs are utilized to classify them with high accuracy. Such models are able to differentiate different species of pests and non-pest organisms. CNNs are especially useful in object identification, image recognition, and classification due to their special quality of learning raw pixel information automatically. CNNs can be trained using large datasets of healthy and diseased plants to distinguish between normal and abnormal conditions of plants, thus enabling the use of the model to detect the presence of pests or diseases early on [50-53]

It is paramount to note that, despite the fact that deep convolutional neural networks (CNNs) are powerful, they require substantial volumes of labelled datasets to be trained, considerable experience, and large amounts of computing power to structure networks and optimize hyperparameters. However, their ability to autonomously derive relevant properties of raw image data has made them a foundation block in many applications of AI, including agriculture and pest control. AgShift has created DeepAgro, a platform using deep learning algorithms to identify pest infestations, disease infections, and other problems based on the images of the crops. A pest-categorisation and identification algorithm that utilises CNNs is the YOLOv3, an AI-driven algorithm. The CNN has taken the image-processing methods that automatically determine the number of insects on the plants. Moreover, using meteorology and computer vision, machine, and deep learning, the platform predicts the pest occurrence (Figure 4) [54-56]. To illustrate, the new YOLOv5 was able to detect pests, including leaf miners, fruit flies, aphids, whiteflies, house flies, and thrips with 96 percent accuracy in greenhouse conditions [57].

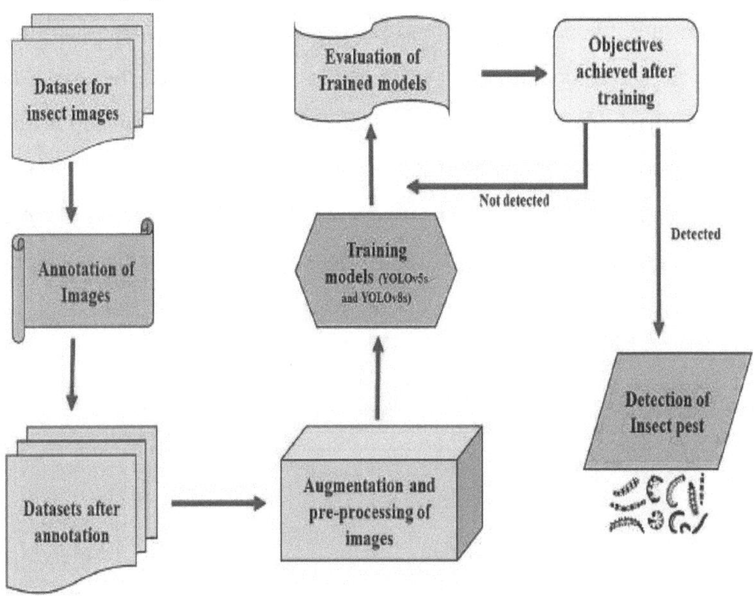

Fig. (4): Using deep learning models, detection and identification of the pest species

The Long Short-Term Memory (LSTM) networks and Recurrent Neural Network (RNNs) can be used to forecast the dynamics of the pest population using historical data, weather conditions, and crop conditions. Moreover, deep learning algorithms are used to process the information about the environment to predict possible pest outbreaks, which can help to take appropriate measures in a timely manner [58].

Deep Learning sensor fusion

Deep learning has the ability to combine the information of various sensors like humidity, temperature and motion sensors, to give a complete picture of the agricultural environment. Devices with several sensors are able to collect information concerning the environment and the presence of pests. Deep-learning models have the ability to combine this information to give correct forecasts of pests outbreaks. To back this up, researchers in the [institution] have created deep-learning inspired models that integrate information collected by weather stations, soil moisture sensors and pest traps to detect pest attacks in vineyards (Fig. 5). An average fruit-fly species, Bactrocera oleae (Rossi), and Ceratitis capitata (Wiedemann) had a top accuracy rate of 93 00 percent in the field with a convolutional neural network model with a camera at a fixed height [59-62].

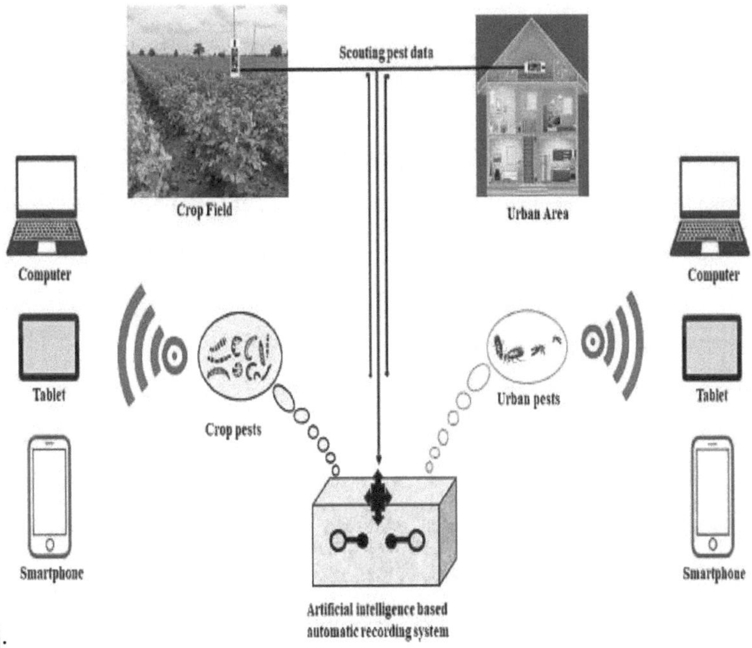

Fig. (5): Surveillance and monitoring of the agricultural crop and urban insect pest using the Internet of Things (IoT)

IoT-based automated pest tracking

Pest sensors and artificial intelligence algorithms installed in IoT devices keep checking the activity of pests in agricultural fields. Those gadgets will have a camera, microphones, and sensors in the environment, which will constantly monitor fields to determine pest activity (Fig. 5). The sensor data can be processed in real time by deep-learning models installed to such devices and alerts can be issued when it detects a pest infestation. They present real-time information about the population of the pests and allow farmers to make informed choices concerning the pest-management plans. Wireless traps with cameras and deep-learning algorithms are formed by the Trapview system developed by EFOS Ltd., which is implemented to track the populations of pests in the orchards and vineyards (Fig. 5). The model used was a multilayer perceptron (MLP), which detected eight hundred pest species with acoustic sensors and PID sensors with an accuracy of 99.78, specificity of 99.64, sensitivity of 99.91, and precision of 99.85, outperforming current models, including YOLOv5, VG-16, DenseNet, and ResNet-50 [6365]. High-level pest detection using deep learning. Deep learning models have the ability to learn intricate patterns and features of large-scale data and can identify pest species in images, DNA, or any other type of data with precision. Deep-learning algorithms can be used to identify pests very fast and with high precision on a field using handheld devices or smartphone apps. Indicatively, the Food and Agriculture Organization (FAO) created the PestNet mobile app that uses deep learning to detect plant pests and diseases using images uploaded by users. The important methods that are used are unsupervised multistage feature learning, scale transferable object detection and Super-Fan which enable object detection in the images [66-68]. Pest detection and monitoring Artificial Intelligence. The AI applications in pest management include a wide range of methods and technologies that can be used to detect, track, and control pests to reduce their effects on agricultural production. The following are some of the major AI applications, Early detection AI systems have the ability to process the data of various sources like satellite shots, unmanned aerial vehicles and surface sensors to detect the initial signs of pests intrusion. Such models employ machine learning to reveal minor patterns and abnormality in the condition of vegetation and inform users about the occurrence of pests or diseases before they manifest themselves [69]. The most regularly used machine-learning algorithms in the farming industry are Decision Tree (DT), Support Vector Machine (SVM), Naive Bayes (NB), and Random Forest (RF) and deep-learning models, including Convolutional Neural Networks (CNN), Deep Convolutional Neural Networks (DCNN), Long Short-Term Memory networks (LSTM) and Deep Belief Networks (DBN) [70].

In recent times, the Tamil Nadu Agricultural University (TNAU) has developed an android based mobile application to detect infestations of the fall armyworm in maize crop under the Integrated Pest Management (IPM) capsules. Using transfer learning and Deep Convolutional Neural Networks (DCNNs), the application reaches high precision in the detection of infected areas with a validation accuracy of 93.47 per cent and a training accuracy of 98.47 per cent. The application allows identifying the presence of fall armyworm infestation early, which can be used via the application of AI algorithms to implement timely intervention measures, i.e., integrated pest management schemes or selective application of pesticides. This computer-based solution can enable farmers to safeguard maize harvests and livelihoods with efficiency and hence highlights the promise of new technologies in solving the issue of crop pests [71]. Precision monitoring The artificial intelligence of drones equipped with cameras with high resolution and sensors can accurately scan forests or farms. To detect the areas of interest, assess crop well-being, and pinpoint pests in imagery, the computer-vision algorithms process the data; this data allow paying attention to specific areas and allocating resources effectively. AI can be used to analyze information about the health of crops, the spread of pests, and the environment to minimize the use of pesticides and decrease the impact on the environment [47]. Grape moths present a threat to the production of wine in vineyards and one solution to the challenge is the application of pheromone traps to monitor and control the pests. Pest-monitoring Implementing cameras, sensors, and AI may be regarded as a nascent trend to increase efficiency. Teixeira et al. [73] consider the segmentation as a pre-processing step to enhance grape-moth recognition in traps images by deep learning. They base their study on the use of U-Net and use ResNet18 and InceptionV3 backbones and test their performance with YOLOv5s and YOLOv8s detectors. Findings show potential segmentation, improving the detection rates by 3 per cent of YOLOv5 and by 1.2 per cent of YOLOv8, which proves the potential of their application in managing plants and preserving grape yields in the vineyard (Fig.). Cotton bollworm insect pests are still not fully controlled, which is why it is important that AI-based pest surveillance networks, using smart traps and AI algorithms, should be able to promote sustainable cotton production, minimize the use of pesticides, and maximize yields. However, AI application in agriculture should be approached with caution to address the possible labour -dismissal issues [72 -74]. Data integration In order to build predictive models used in pest forecasting, AI combines many types of data, such as meteorological patterns, agricultural phenology, and pest life-cycles information. These models make use of past data as well as real time data to forecast pest outbreaks hence providing proactive management strategies and responsive measures [6]. Data of insect numbers, movements and behavior stipulated by IoT cameras and sensors are utilized in accurate and timely recognition and tracking of the presence of pests and diseases [75]. The complexity of pest development, environment factors and pesticide overuse is faced by farmers. To solve these problems, scholars have come up with an IPM system that utilizes AI, data analytics, and various sources of data. The proposed innovative solution aims at streamlining pest-management activities in the agricultural sector by providing farmers with a way of more effective and sustainable methods of crop protection and reducing the environmental harm and the emergence of pesticide resistance [75, 76].

Automation AI-powered devices automate the monitoring and identification of pests, results in decreasing need of labour and enhancing the crop productivity. [50]. Rustia et al. [78] developed an automatic method for recognizing minute insect pests of greenhouse on sticky paper trap images.. The GoogleNet employed with CNN model detected the plant pests 93.78 % accurately, this helps the farmers to adopt the smartphone application for identification of the pests and disease encountering on their crop fields [77, 79].

Decision making activities

They are deployed for tasks like observation, data collection and analysis and disaster management. Drones employ different techniques for plant spraying, including gaseous, kinetic, centrifugal, or hydraulic methods [81]. They also help farmers to monitor sunlight exposure and identify areas for targeted pest spraying using ML (Fig. 6). Chen et al. [82] utilizes edge intelligence for a smart pest recognition system. The system then devises efficient pesticide spraying routes for an agricultural drone using pest positions. There are several decision tools (DTs) available such as decision rules, risk algorithms, decision support systems and intervention threshold which are made available for farmers and advisors via internet based system [83] to make timely decision in crop cultivation.

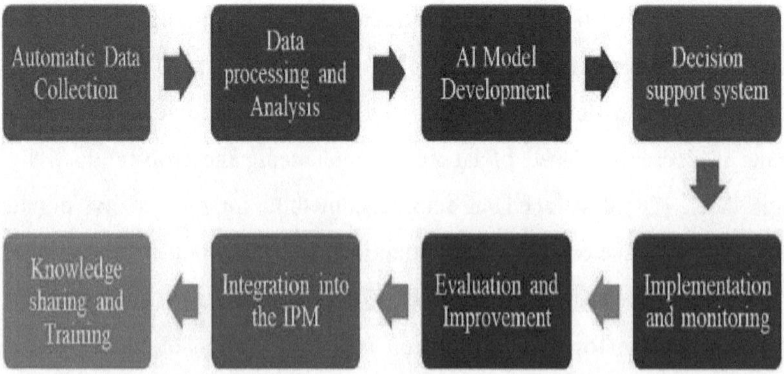

Fig. (6): Better utilizing and adoption of edge intelligence technique for a smart pest and weather recognition system for sustainable crop cultivation.

Sensors in artificial intelligence

Sensors, as defined by Bannerjee et al. [19], are devices that convert the results of chemical or physical reactions in living organisms or the environment into useful signals that can be analyzed using specialized programs. They gather information in different ways: cameras take pictures, microphones capture sound, LiDAR and radar sensors use signals to measure distance and IMUs track movement [84]. ML is often used to teach AI systems how to interpret sensor data and make decisions based on it [77]. In some cases, sensors are part of a feedback loop, where their readings influence the actions of the system and because many AI tasks need to happen quickly, sensors often work in real-time to provide rapid feedback to the AI system. [63].

Automatic pest detection using sensor

Employing data mining techniques to pinpoint important influences on Helicoverpa armigera (Hübner) infestation. The researchers examined season, cotton growth stage, and abiotic variables like minimum and maximum temperature, precipitation, morning and evening relative humidity, and the number of rainy days per week among other factors. Their findings showed that the development stage of cotton was the most important element affecting the incidence of the insect pest [95]; other aspects included humidity, number of rainy days, and temperature. Likewise, spectroradiometer used to evaluate Thrips tobaci (Lind) damage on cotton measured the canopy reflectance from 70 to 90 days following seeding. Red band and green-red vegetative index were sensitive to thrips damage, with a characteristic peak in the blue region [96]

Remote sensing using a spectroradiometer allowed Nigam et al. [97] to evaluate whitefly infestation and biotic stress. Using AI sensors, researchers have come up with creative means to deal with insect pests in farming.. In orchards also, clever insect traps armed with AI sensors have been installed to give real-time information on pest numbers [1]. (Fig.7). Drone-based surveillance systems have also been used to monitor agricultural areas for pests on a wide scale. AI algorithms are used to evaluate aerial imagery taken by drones fitted with LiDAR sensors and multispectral cameras in order to find indications of pest infestations [63].

An insect signal analysis program called "Digitize, Analyze, View, Insect Sounds" (DAVIS), in conjunction with microphones and amplifiers, was able to identify the larval outbreaks of Anoplophora glabripennis (Motschulsky) with 90% accuracy and Rhynchophorus ferrugineus (Olivier) with 80% accuracy [98–100].

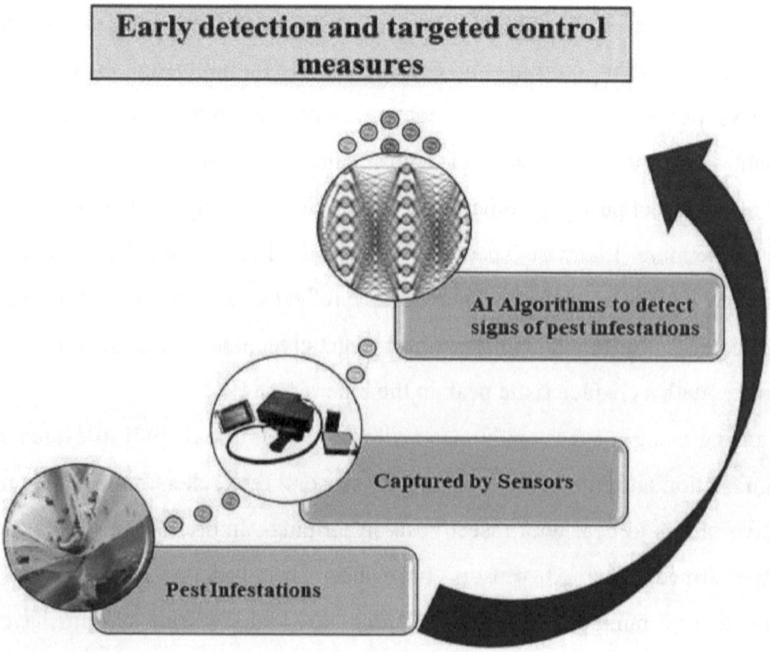

Fig. (7): Use of Artificial intelligence based sensors in sustainable and environmentally friendly pest management practices.

Conclusion

The potential for increasing the efficacy and efficiency of pest control methods is substantial with regard to algorithms for pest management. Adopting the bottleneck will transform traditional farming into automated, sustainable farming.

References

1. N. Ahmed, D. De, I. Hussain Internet of Things (IoT) for smart precision agriculture and farming in rural areas IEEE Internet Things J., 5 (6) (2018), pp. 4890-4899

2. C.M. Oliveira, A.M. Auad, S.M. Mendes, M.R. Frizzas Crop losses and the economic impact of insect pests on Brazilian agriculture Crop Prot., 56 (2014), pp. 50-54
3. W. Ding, G. Taylor Automatic moth detection from trap images for pest management Comput. Electron. Agric., 123 (2016), pp. 17-28
4. W. Li, D. Wang, M. Li, Y. Gao, J. Wu, X. Yang Field detection of tiny pests from sticky trap images using deep learning in agricultural greenhouse Comput. Electron. Agric. (2021), p. 183
5. Y. Sun, X. Liu, M. Yuan, L. Ren, J. Wang, Z. Chen Automatic in-trap pest detection using deep learning for pheromone-based dendroctonus valens monitoring Biosys. Eng., 176 (2018), pp. 140-150
6. R. Dutta, S. Sharma Artificial Intelligence for Sustainable Agriculture: A Comprehensive Review IEEE Access, 8 (2020), pp. 73641-73672
7. M. Preti, F. Verheggen, S. Angeli Insect pest monitoring with camera-equipped traps: strengths and limitations J. Pest Sci., 94 (2021), pp. 203-217, 10.1007/s10340-020-01309-4
8. J. Streich, J. Romero, J.G.F.M. Gazolla, D. Kainer, A. Cliff, E.T. Prates, J.B. Brown, S. Khoury, G.A. Tuskan, M. Garvin, D. Jacobson Can exascale computing and explainable artificial intelligence applied to plant biology deliver on the United Nations sustainable development goals Curr. Opin. Biotechnol., 61 (2020), pp. 217-225
9. V Partel, S Charan Kakarla, Y Ampatzidis Development and evaluation of a low-cost and smart technology for precision weed management utilizing Artificial Intelligence Comput. Electron. Agric., 157 (2019), pp. 339-350
10. Z. Iqbal, M.A. Khan, M. Sharif, J.H. Shah, M.H. URRehman, K. Javed An automated detection and classification of citrus plant diseases using image

processing techniques: A review Comput. Electron. Agric., 153 (2018), pp. 12-32

11. D.I. Patrício, R. Rieder Computer vision and Artificial Intelligence in precision agriculture for grain crops: A systematic review Comput. Electron. Agric., 153 (2018), pp. 69-81
12. J. Doe, A. Smith Applications of artificial intelligence in pest management: detection and identification techniques J. Pest Manag., 15 (2) (2023), pp. 78-89
13. M.K. Hanif, S.Z. Khan, M. Bibi Applications of artificial intelligence in pest management Artificial Intelligence and Smart Agriculture Applications, Auerbach Publications (2022), pp. 277-300
14. A. Uzhinskiy Advanced Technologies and Artificial Intelligence in Agriculture Applied Math, 3 (4) (2023), pp. 799-813
15. D.K. Pandey, R. Mishra Towards sustainable agriculture: Harnessing AI for global food security Artif. Intell. Agric. (2024), 10.1016/j.aiia.2024.04.003
16. S. Kanwal, M.A. Khan, S. Saleem, M.N. Tahir, S.T. Muntaha, T. Samreen, S . Javed, M.Z. Nazir, B. Shahzad Integration of precision agriculture techniques for pest management Environ. Sci. Proc., 23 (1) (2022), p. 19
17. J.H. Zhang, F.T. Kong, J.Z. Wu, S.Q. Han, Z.F. Zhai Automatic image segmentation method for cotton leaves with disease under natural environment J. Integr. Agric., 17 (8) (2018), pp. 1800-1814
18. S. Russell, P. Norvig Artificial Intelligence: A modern approach Pearson, Harlow (2022) global edition
19. G. Bannerjee, U. Sarkar, S. Das, I. Ghosh Artificial Intelligence in agriculture: a literature survey Int. J. Sci. Res. Comp. Sci. Appl. Manag Stud., 7 (3) (2018), pp. 1-6

20.X. Liang Few-shot cotton leaf spots disease classification based on metric learning Plant Methods, 17 (2021), pp. 1-11

21.Goodfellow, Y. Bengio, A. Courville Sequence modeling: recurrent and recursive nets Deep learning (2016), pp. 367-415

22.S. Kar, K. Nagasubramanian, D. Elango, M.E. Carroll, C.A. Abel, A. Nair, D.S. Mueller, M.E. O'Neal, A.K. Singh, S. Sarkar, B. Ganapathysubramania n Self-supervised learning improves classification of agriculturally important insect pests in plants Plant Phenome J., 6 (1) (2023), p. e20079

23.C.M. Bishop, N.M. Nasrabadi Pattern recognition and machine learning, 4, springer, New York (2006), p. 738

24.J.G.A. Barbedo Using digital image processing for counting whiteflies on soybean leaves J. Asia Pac. Entomol., 17 (4) (2014), pp. 685-694

25.Geron, A. (2019). Hands-On Machine Learning with Scikit-Learn, Keras, and TensorFlow (2nd ed.). OReilly Media.

26.C. Hernandezcastillo, H.H. Guedeanoriega, M.A. Rodriguezgarcia, F. Garcia sanchez Pest recognition using natural language processing International Conference on Technologies and Innovation, Springer International Publishing, Cham (2019), pp. 3-16

27.ME Karar, F Alsunaydi, S Albusaymi, S Alotaibi A new mobile application of agricultural pests recognition using deep learning in cloud computing system Alex. Eng. J., 60 (5) (2021), pp. 4423-4432

28.D Motka, R Badaró, A Santos, F. Kirchner Use of Artificial Intelligence on the Control of VectorBorne Diseases Vectors and Vector-Borne Zoonotic Diseases (2018), p. 439

29.Y Bouroubi, P Bugnet, T NguyenXuan, C Gosselin, C Bélec, L Longchamps , P. Vigneault Pest detection on UAV Imagery using a deep convolutional neural network Proceedings of International Society of Precision Agriculture (2018), pp. 1-11

30. L. Deng, Y. Wang, Z. Han, R. Yu Research on insect pest image detection and recognition based on bio-inspired methods Biosystems Eng., 169 (2018), pp. 139-148
31. M. Khanramaki, E.A. Asli-Ardeh, E. Kozegar Citrus pests classification using an ensemble of deep learning models Comput. Electron. Agric., 186 (2021), Article 106192
32. Penn State University, 2024. Plant Village, Available at: https://plantvillage.psu.edu/, Accessed on: 17/05/2024.
33. A.N. Alves, W.S. Souza, D.L. Borges Cotton pests classification in field-based images using deep residual networks Comput. Electron. Agric., 174 (2020), Article 105488
34. T. Kasinathan, D. Singaraju, S.R. Uyyala Insect classification and detection in field crops using modern machine learning techniques Inf. Process. Agric., 8 (3) (2021), pp. 446-457
35. G. Capizzi, G.R.A.Z.I.A. LoSciuto, C. Napoli, E. Tramontana, M.A.R.C.I.N. Woźniak A novel neural networks-based texture image processing algorithm for orange defects classification Int. J Adv. Comput. Sci. Appl., 13 (2) (2016), pp. 45-60
36. Spornado, 2024. Spornado Sampler, Available at: https://spornadosampler.com/, Accessed on: 17/05/2024.
37. R. Badia-Melis, J.P. Qian, B.L. Fan, P. Hoyos-Echevarria, L. Ruiz-García, X.T. Yang Artificial neural networks and thermal image for temperature prediction in apples Food Bioprocess Technol., 9 (2016), pp. 1089-1099

38.EFOS Ltd, 2024. Trapview, Available at: https://trapview.com/, Accessed on: 17/05/2024.

39.K.U. Singh, A. Kumar, L. Raja, V. Kumar, A.K. Singhkushwaha, N. Vashne y, M. Chhetri An artificial neural network-based pest identification and control in smart agriculture using wireless sensor networks J. Food Qual. (2022), pp. 1-12

40.Liu B, Hu Z, Zhao Y, Bai Y, Wang Y. Recognition of Pyralidae Insects Using Intelligent Monitoring Autonomous Robot Vehicle in Natural Farm Scene. arXiv preprint ar. 2019; Xiv.1903:10827.

41.P Gao, Y Zhang, L Zhang, R Noguchi, T. Ahamed Development of a recognition system for spraying areas from unmanned aerial vehicles using a machine learning approach Sensors, 19 (2) (2019), p. 313

42.University of Georgia, 2024. PestID, Available at: https://pestid.com.au/, Accessed on: 17/05/2024.

43.B.J. Spiesman, C. Gratton, R. Hatfield, W.H. Hsu, S.J. Jepsen, B.P McCorna ck, K. Patel, G. Wang Assessing the potential for deep learning and computer vision to identify bumble bee species from images Sci. Rep., 11 (2021), p. 7580

44.S. Huang, M. Zhou, G. Geng An angle-independent Gabor-SVM insect recognition J. Chinese Computer Syst., 31 (2010), pp. 143-146

45.Y. Dong, F. Xu, L. Liu, X. Du, H. Ye, W. Huang, Y. ZhuMonitoring and forecasting for disease and pest in crop based on WebGIS system 2019 8th International Conference on Agro-Geoinformatics (Agro-Geoinformatics), IEEE (2019), pp. 1-5

46.Palani H. K., Ilangovan S., Senthilvel P. G., Thirupurasundari D. R. and Rajesh Kumar K, 2023. AI-powered predictive analysis for pest and disease forecasting in Crops. *International Conference on Communication, Security and Artificial Intelligence (ICCSAI)*, Greater Noida, India, 2023, pp. 950-954.

47.D. Markovic, D. Vujicic, S. Tanaskovic, B. Dordevic, S. Randic, Z. Stamenk ovic Prediction of pest insect appearance using sensors and Machine Learning Sensors, 21 (14) (2021), p. 4846

48.CSIRO, 2024. RapidAIM digital crop defence, Available at: https://www.csiro.au/en/research/animals/pests/rapidaim-digital-crop-defence, Accessed on: 15/05/2024.

49.A.A. Alfarisy, Q. Chen, M. Guo Deep learning-based classification for paddy pests & diseases recognition Proceedings of 2018 international conference on mathematics and Artificial Intelligence (2018), pp. 21-25

50.D. Popescu, A. Dinca, L. Ichim, N. Angelescu New trends in detection of harmful insects and pests in modern agriculture using artificial neural networks. a review Front. Plant Sci., 14 (2023), Article 1268167

51.D.J. Patel, N. Bhatt Insect identification among deep learning's metaarchitectures using TensorFlow Int. J. Eng. Adv. Technol., 9 (2019), pp. 1910-1914, 10.35940/ijeat.A1031.109119

52.J. Liu, X. Wang Tomato diseases and pests detection based on improved Yolo V3 convolutional neural network Front. Plant Sci., 11 (2020), Article 521544

53.Y. Li, H. Wang, L.M. Dang, A. Sadeghi-Niaraki, H. Moon Crop pest recognition in natural scenes using convolutional neural networks Comput. Electron. Agric., 169 (2020), Article 105174

54.A.E. Douglas Strategies for enhanced crop resistance to insect pests Annu. Rev. Plant Biol., 69 (2018), pp. 637-660

55.M.K. Khan, M.O. Ullah Deep transfer learning inspired automatic insect pest recognition Proceedings of the 3rd International Conference on Computational Sciences and Technologies, Mehran University of Engineering and Technology, Jamshoro, Pakistan (2022), pp. 17-19
56.C. Li, T. Zhen, Z. Li Image classification of pests with residual neural network based on transfer learning Appl. Sci., 12 (9) (2022), p. 4356
57.X. Zhang, J. Bu, X. Zhou, X. Wang Automatic pest identification system in the greenhouse based on deep learning and machine vision Front. Plant Sci., 14 (2023), Article 1255719
58.J. Parraga-Alava, R. Alcivar-Cevallos, J.A. Riascos, M.A. Becerra Aphids detection on lemons leaf image using convolutional neural networks M. Botto-Tobar (Ed.), Systems and information sciences: proceedings of ICCIS 2020, Springer International Publishing AG, Cham (2021), pp. 16-27
59.N.E.M. Khalifa, M.O.H.A.M.E.D. Loey, M.H.N. Taha Insect pests recognition based on deep transfer learning models J. Theor. Appl. Inf. Technol, 98 (1) (2020), pp. 60-68University of California, Davis research: UC Davis News.
60.R.D. Clark Putting deep learning in perspective for pest management scientists Pest Manage. Sci., 76 (7) (2020), pp. 2267-2275
61.M Tannous, C Stefanini, D. Romano A deep-learning-based detection approach for the identification of insect species of economic importance Insects, 14 (2) (2023), p. 148
62.A. Khattab, S.E. Habib, H. Ismail, S. Zayan, Y. Fahmy, M.M. Khairy An IoT-based cognitive monitoring system for early plant disease forecast Comput. Electron. Agric., 166 (2019), Article 105028

63.E. Omia, H. Bae, E. Park, M.S. Kim, I. Baek, I. Kabenge, B.K. Cho Remote sensing in field crop monitoring: a comprehensive review of sensor systems, data analyses and recent advances Remote Sensing, 15 (2) (2023), p. 354

64.RK Dhanaraj, M.A Ali Deep Learning-Enabled Pest Detection System Using Sound Analytics in the Internet of Agricultural Things Engineering Proceedings, 58 (1) (2023), p. 123

65.I. Goodfellow, Y. Bengio, A. Courville Deep Learning MIT Press (2016)

66.FAO, 2024. PestNet., Available at: https://www.plantwise.org/pestnet, Accessed on: 17/05/2024.

67.B. Prasath, M. Akila IoT-based pest detection and classification using deep features with enhanced deep learning strategies Eng. Appl. Artif. Intell., 121 (2023), Article 105985

68.V.K. Singh, D. Jat AI in Agriculture: A Comprehensive Review Comput. Electron. Agric., 183 (2021), Article 106015

69.M Chithambarathanu, MK. Jeyakumar Survey on crop pest detection using deep learning and machine learning approaches Multimed Tools Appl., 11 (2023), pp. 1-34

70.R. Prabha, J.S. Kennedy, G. Vanitha, N. Sathiah, M.B. Priya Android application development for identifying maize infested with fall armyworms with Tamil Nadu agricultural university integrated proposed pest management (TNAU IPM) capsules J. Appl Nat. Sci., 14 (SI) (2022), pp. 138-144

71.S.P. Mohanty, D.P. Hughes, M. Salathe Using deep learning for image-based plant disease detection Front. Plant Sci., 7 (2016), Article 215232

72.A.C. Teixeira, G.A. Carneiro, R. Morais, J.J. Sousa, A. Cunha Segmentation as a Pre-processing for Automatic Grape Moths Detection EPIA Conference on Artificial Intelligence, Springer Nature Switzerland, Cham (2023), pp. 388-398

73. U. Ghate, H. Kulkarni Artificial Intelligence & Nature-Based Solutions in Agriculture: A BT Cotton Pest Management Case Study in India Qeios (2024)
74. I. Samal, A.K. Pradhan, T.K. Bhoi Role of Artificial Intelligence in Integrated Pest Management AgroScience Today, 4 (5) (2023), pp. 0579-0586
75. J.P. Deguine, J.N. Aubertot, R.J. Flor, F. Lescourret, K.A. Wyckhuys, A. Ra tnadass Integrated pest management: good intentions, hard realities. a review Agron. Sustainable Dev., 41 (3) (2021), p. 38
76. V.S. Chandra Role of artificial intelligence in Indian agriculture: A review Agr. Rev., 44 (4) (2023), pp. 558-562
77. D.J.A. Rustia, J.J. Chao, L.Y. Chiu, Y.F. Wu, J.Y. Chung, J.C. Hsu, T.T. Lin Automatic greenhouse insect pest detection and recognition based on a cascaded Deep Learning classification method J. Appl. Entomol., 145 (3) (2021), pp. 206-222
78. I.N. Yulita, M.F.R. Rambe, A Sholahuddin, A.S. Prabuwono A convolutional neural network algorithm for pest detection using GoogleNet AgriEngineering, 5 (4) (2023), pp. 2366-2380
79. T.A. Shaikh, T. Rasool, F.R. Lone Towards leveraging the role of machine learning and artificial intelligence in precision agriculture and smart farming Comput. Electron. Agric., 198 (2022), Article 107119
80. Kumar, N., 2023. Leveraging artificial intelligence in agriculture: transforming the future of farming. Illumenia.

81.C.J. Chen, Y.Y. Huang, Y.S. Li, Y.C. Chen, C.Y. Chang, Y.M. Huang Identification of fruit tree pests with deep learning on embedded drone to achieve accurate pesticide spraying IEEE Access, 9 (2021), pp. 21986-21997
82.V. Rossi, T. Caffi, I. Salotti, G. Fedele Sharing decision-making tools for pest management may foster implementation of Integrated Pest Management Food Security, 15 (6) (2023), pp. 1459-1474
83.R. ToscanoMiranda, M. Toro, J. Aguilar, M. Caro, A. Marulanda, A. Trebilc ok Artificial-Intelligence and sensing techniques for the management of insect pests and diseases in cotton: a systematic literature review J.Agric. Sci., 160 (1-2) (2022), pp. 16-31
84.S MacDougall, F Bayansal, A. Ahmadi Emerging methods of monitoring volatile organic compounds for detection of plant pests and disease Biosensors, 12 (4) (2022), p. 239
85.S. Xu, Z. Zhou, H. Lu, X. Luo, Y. Lan, Y. Zhang, Y. Li Estimation of the age and amount of brown rice plant hoppers based on bionic electronic nose use Sensors, 14 (2014), pp. 18114-18130
86.A. Rizzolo, G. Bianchi, P. Lucido, B. Cangelosi, L. Pozzi, G. Villa, F. Clem atis, C. Pasini, P. Curir Electronic nose for the early detection of red palm weevil (rhynchophorus ferrugineous olivier) infestation in palms: Preliminary results Acta Hortic., 1099 (2015), pp. 347-356
87.M.F. Rutolo, D. Iliescu, J.P. Clarkson, J.A. Covington Early identification of potato storage disease using an array of metal-oxide based gas sensors Postharvest Biol. Technol., 116 (2016), pp. 50-58
88.T. Wen, L. Zheng, S. Dong, Z. Gong, M. Sang, X. Long, M. Luo, H. Peng Rapid detection and classification of citrus fruits infestation by Bactrocera

dorsalis (Hendel) based on electronic nose Postharvest Biol. Technol., 147 (2019), pp. 156-165

89. S. Hazaika, R. Choudhury, S. Saikia, U. Sarma Pathogen Detection in Khasi Mandarin Orange using Serological and Electronic Nose Diagnostic Technique Int. J. Innov. Technol. Explor. Eng., 9 (2020), pp. 2981-2985
90. B.D. Lampson, D.C. Degenhardt, J.K. Greene, A. Khalilian, Y.J. Han Development of a portable electronic sensor for detection of the kudzu bug, megacopta cribraria (Fabricius) (Hemiptera: Plataspidae) Adv. Entomol., 05 (2017), pp. 75-86
91. C. Li, R. Gitaitis, B. Tollner, P. Sumner, D. MacLean Onion sour skin detection using a gas sensor array and support vector machine Sens. Instrum. Food Qual. Saf., 3 (2009), pp. 193-202
92. W.G. Henderson, A. Khalilian, Y.J. Han, J.K. Greene, D.C. Degenhardt Detecting stink bugs/damage in cotton utilizing a portable electronic nose Comput. Electron. Agric., 70 (2010), pp. 157-162
93. C. Li, N.E. Schmidt, R. Gitaitis Detection of onion postharvest diseases by analyses of headspace volatiles using a gas sensor array and GC-MS LWT-Food Sci. Technol., 44 (2011), pp. 1019-1025
94. M. Pratheepa, A Verghese, H. Bheemanna Shannon information theory a useful tool for detecting significant abiotic factors influencing the population dynamics of *Helicoverpa armigera* (Hubner) on cotton crops Ecol. Modell., 337 (2016), pp. 25-28
95. G. Ranjitha, M.R. Srinivasan, A. Rajesh Detection and estimation of damage caused by thrips *Thrips tabaci* (Lind) of cotton using hyperspectral radiometer Agrotechnology, 3 (1) (2014), p. 123

96. R. Nigam, R. Kot, S.S. Sandhu, B.K. Bhattacharya, R.S. Chandi, M. Singh, J . Singh, K.R. Manjunath Ground-based hyperspectral remote sensing to discriminate biotic stress in cotton crop Multispectral, Hyperspectral, and Ultraspectral Remote Sensing Technology, Techniques and Applications VI, 9880, SPIE (2016), pp. 89-98

97. I. Potamitis, I. Rigakis, N.A. Tatlas Automated surveillance of fruit flies Sensors, 17 (2017), p. 110

98. R.W. Mankin, M.T. Smith, J.M. Tropp, E.B. Atkinson, D.Y. Jong Detection of *Anoplophora glabripennis* (Coleoptera: Cerambycidae) larvae in different host trees and tissues by automated analyses of sound-impulse frequency and temporal patterns J. Econ. Entomol., 101 (2008), pp. 838-849

99. N.J. Herrick, R.W. Mankin Acoustical detection of early instar *Rhynchophorus ferrugineus* (Coleoptera: Curculionidae) in canary island date palm, phoenix canariensis (Arecales: Arecaceae) Fla. Entomol., 95 (2012), pp. 983-990

100. A.N. Brezolin, J. Martinazzo, D.K. Muenchen, A.M. De Cezaro, A.A. Rigo, C. Steffens, J. Steffens, M.C. Blassioli-Moraes, M. Borges Tools for detecting insect semiochemicals: a review Anal. Bioanal.Chem., 410 (17) (2018), pp. 4091-4108

Chapter (2): AI and IoT-powered for crop pest and disease detection

Mohamed Abdel-Raheem

Pests & Plant Protection Department, Agricultural and Biological Research Institute, National Research Centre, 33rd El Buhouth St, (Postal code: 12622) Dokki, Giza, Egypt.

E-mail: abdelraheem_nrc@yahoo.com, abdelraheem_nrc@hotmail.com

Tel: (+2)01155527583, (+2)01009580797

Abstract:

The use of Artificial Intelligence (AI) has revolutionized the agricultural sector, making it more resilient, sustainable and productive. The main system functions comprise recurrent neural networks (RNNs) time-series analysis and convolutional neural networks (CNNs) high-resolution satellite image processing, among other machine-learning-based disease and pests detection functions. The predictive performance of the model has been critically tested on past epidemics, and it is significantly better than traditional mechanisms. This method will allow farmers to get a warning in time and consequently reduce the effects of biotic risks. The predictive insights also help agronomists and policymakers to maximise pest and disease management, hence promoting sustainable agriculture and food security. Finally, predictive analysis, implemented by AI, has a chance to transform the sphere of controlling pests and diseases in agriculture. The given research demonstrates how AI can help to revolutionize crop harvests by combining various pieces of data and cutting-edge algorithms. Climate change adds to the struggles of ensuring the health of crops by impacting invasive pests and disease outbreaks especially in cereal crops hence causing a significant loss in yield. Therefore, new technologies are needed to track the health of crops during the initial phase of development to the harvest time. Despite the use of diverse technologies, such as the Internet of Things (IoT), machine learning (ML) and artificial intelligence (AI), portable solutions, resource-efficient, and energy-saving solutions are needed, which would fit in resource-limited environments like edge agriculture applications.

Key words: AI, IoT-powered, crop pest, disease detection

Introduction

Agriculture is an important sector of the world economy and it has been instrumental in the food security and rural growth, especially in the developing nations. Maize, wheat, and rice are areas of the cereal which are major sources of food that are foodstuff for sustaining the economies of many African countries and their subsistence food. Their production is however, being threatened by invasive pests and devastating diseases. Further, factors that lead to climate changes such as changes in temperature condition and erratic weather condition have significant contribution to the establishment of favourable conditions which favour the outbreaks of pests and diseases. Therefore, new technologies need to be used in a sustainable solution[1-5]. Conventional ways of managing outbreak of pests and diseases are dependent on manual inspection and extensive use of chemical pesticides all over the fields by farmers or professionals. Although these are effective to some extent, they are time consuming, labour intensive and subject to human error, which results in delayed response, and large-scale destruction of crops. Moreover, introduction of the integrated pest management (IPM) programmes to promote sustainable pest management practices remains reliant on human knowhow, and is yet to be real-time automated and induce resistance of pests to pesticides. These shortcomings of the traditional approaches demonstrate the necessity to find new technologies to resolve these issues[6-10]. The Internet of Things (IoT) is an emerging technology that allows the virtual interaction of physical objects. Precision agriculture is a technological instrument that is using this technology to support remote and real-time tracking of data to make data-driven decisions. There are different agricultural sensors, actuators, and processing devices that capture essential agro-environmental measurements that are soil parameters, crop health status, and weather patterns to monitor and make automated decisions in real-time. In managing pests and diseases, IoT-driven devices can detect and track the initial signs of infestations by constant data collection and analysis of information in the field which minimise the use of manual inspection and optimise resource utilisation. Nevertheless, much of the currently available IoT solutions relies on cloud computing which is characterised by latency, bandwidth and connectivity issues among smallholder farmers with limited access to accessible and reliable internet infrastructure, especially in rural/ remote regions. Thus, edge computing has been proposed to overcome these constraints and run data on IoT devices, which makes use of cloud-based solutions less critical and responsive in real time [11-17]. Precision Agriculture has resulted in the creation of superior smart solutions through artificial intelligence (AI), machine learning (ML) and the Internet of Things (IoT) to control and monitor pest and disease outbreaks via real-time decision-making. Convolutional neural networks (CNNs) have proven to be incredibly good at processing images, and so they can detect a pest or a disease outbreak at an early stage and stop it spreading to large regions. These capabilities have been improved by cloud-based satellite and drone-based imaging systems which have been capable of covering large areas and even finding hard-to-reach areas. However, the technologies are costly, computationally high and demand high-level infrastructure development. Furthermore, the majority of these technologies are based on cloud computing and need to have good internet connectivity, which makes it impossible to use in rural and remote regions with low internet connectivity. To address such barriers, lightweight AI systems combined with inexpensive and small edge device are required in precision agriculture to benefit smallholder farmers [18-25].

To solve these issues, this paper presents a low power portable Internet-of-Things device, which integrates a lightweight convolutional neural network model, which is designated Tiny-LiteNet. The product is specifically designed when it comes to edge computing in precision agriculture. It has a high-definition camera to capture images, a Raspberry 5 to process the image on-board and a GSM/GPRS to enable communication. Tiny-LiteNet is specially designed to work on edge computing of images in real-time, so that it can be used to monitor without the need of having powerful internet connectivity. The experimental findings indicate that Tiny-LiteNet has a high level of performance efficiency with an accuracy score of 98.6, a smaller model size of 1.2MB, and 1.48 million parameters compared to all the conventional deep-learning models: VGGNet-16, Inception3, ResNet50, DenseNet121, EfficientNetB0, and MobileNetv2. The small power usage and usability of the device combined with minimal reliance on high-speed internet makes it a viable activity in any smallholder farm, enabling real-time monitoring and sustainable agriculture. The main works of this paper include: (1) the creation of a small AI-IoT edge device, portable and low-power, to operate in real-time to detect pests and diseases; (2) the design and creation of an effective lightweight CNN model, Tiny-LiteNet, that could be deployed on edge devices with limited resources; and (3) the demonstration of the effectiveness of this model in smallholder farming conditions with limited connectivity. Agricultural pests and diseases IoT technology has been found as one of the real-time solutions to monitor and control pests and diseases. In [26], a CROPCARE system was designed to control crop diseases in real time by combining the IoT with mobile vision, based on MobileNetv2 with impressive results. In reference [27], the parallel and distributed simulation framework (PDSF) combined with the IoT was suggested to monitor pests. This computing system uses a central graphical processing unit (GPU) to control the amount of connected sensors, hence, controlling and monitoring crop pests with an accuracy of 98.65. A smart pest detection and management system was created on cotton plants in reference[29], where motion-detection sensors were used to detect pests and an automated targeted spray was delivered through drone. However, a majority of the current IoT solutions are based on cloud computing, which is not a feasible solution in rural regions with low internet access. The use of machine-learning algorithms, especially deep learning, has actually transformed the practice in agriculture by automating some of the labour-intensive work, such as the pest and disease detection [30,31]. Machine-learning prediction opportunities allow detecting outbreaks of pests and diseases at an early stage before they get out of control. N. Ullah et al. [32] introduced DeepPestNet; a model of convolutional neural network that recognised and classified 10 pest species with an accuracy of 98.92 per cent. In the same way, an enhanced EfficientNetV2-B4 model has been suggested by W.Albattah et al. [33] using the customization of the existing EfficientNet architecture with average precision, recalls, and accuracy of 99.63-B4, 99.93-B4 and 99.99-B4 respectively. Moreover, the article in reference [34] suggested a deep-learning model, which used the Faster R CNN architecture along with a mobile application to identify and localise pests on images with a stunning accuracy of 99.0 -1. The use of unmanned aerial vehicles (UAVs), artificial intelligence, and the Internet of Things has contributed to the development of precision agriculture and accurate monitoring of plant-health. Real-time data collection, analysis, prediction, and informed decision-making using the integration of these technologies is used to improve the accuracy and responsiveness of pest and disease management systems [35]. Crop diseases and pest infestations were identified and tracked with the help of a UAV with deep learning with the help of the YOLOv5 model in a study that was carried out in reference[36], and the results were impressive with the highest average precisions of 96.0, greatest recall of 93.0, and average precisions of 95.0. Also, an example is an invasive pest R. Fu [37], who proposed a swarm system of drones with a CNN model based on YOLOv10 to detect and geo-localise invasive pests like Araujia sericifera and Cortaderia selloana in an organic orange grove with an F1 score up to 80. Though such impressive success has been achieved, the majority of AI, ML, and IoT applications focus on high-resource environments, thus leaving an unmet need of affordable, scalable, and energy-efficient solutions to target smallholder farmers in the developing world [38, 39]. As Figure 1 shows, the system starts with the acquisition of the images of plants in the field and pre-processing them to be analyzed in a more effective way by means of the Tiny-LiteNet model. When a pest or disease is detected, the system triggers a notification system by notifying the farmers at the location and sending a notification message with a description of the plant-health status. In addition, the photos of the identified plant-health states are saved on the machine and sent to a cloud storage system as soon as the internet connection is established to be analyzed.

Fig. 1

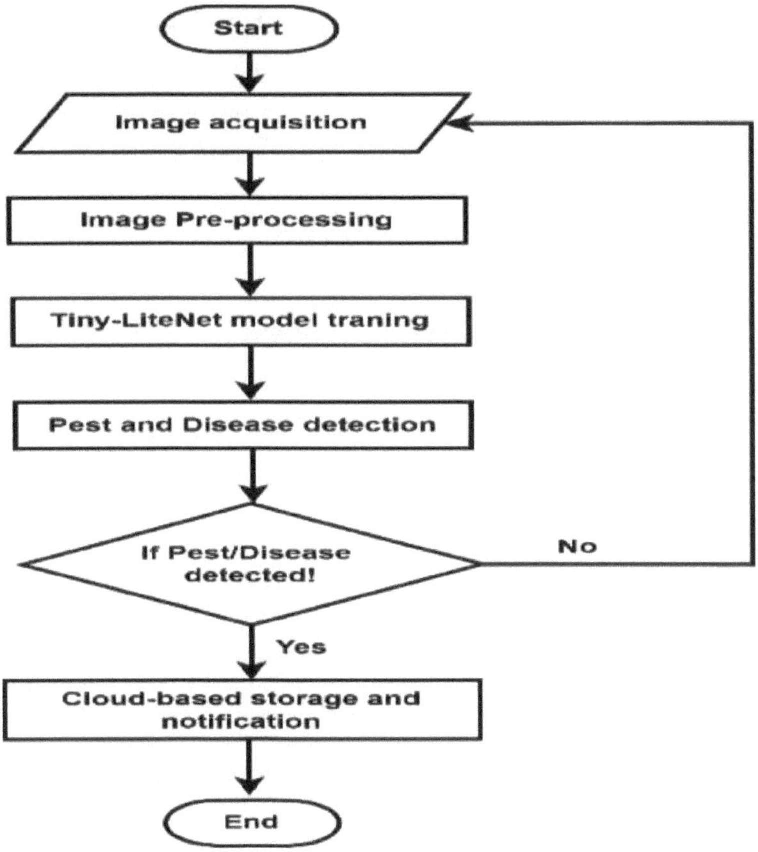

The flowchart of the system

The suggested system design of the AI-driven portable edge device (with a high-definition camera to capture real-time crop images) is presented in Figure 2 and includes a Raspberry 5 to conduct on-board processing and a GSM/GPRS module to network the edge device with the cloud. In addition, it contains Tiny-LiteNet a lightweight convolutional neural network model based on MobileNetV2, but optimized to run on the edge and specifically to detect crop pests and diseases.

Fig. 2

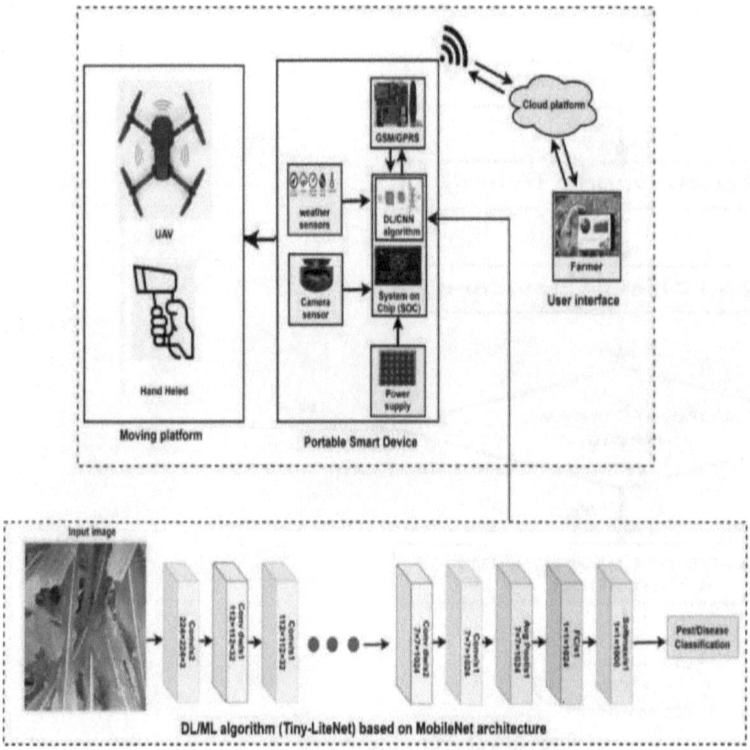

Overall system architecture

MinuteNet Model Development. In Figure 3, a pretrained MobileNetV2 model was used as the teacher model, and it is out of which basic knowledge was learned.

Also, a six squeeze excitation depth wise block student model was used to include better representation of features, lower computational cost and better model accuracy; this was augmented with a fully connected layer to transform the high level feature maps into class scores to use in classification tasks. The overall architecture makes the MobileNetV2 architecture much cheaper to compute but has a relative accuracy reduction of 1% which has no overall negative impact on the pest and disease detection performance.

Fig. 3

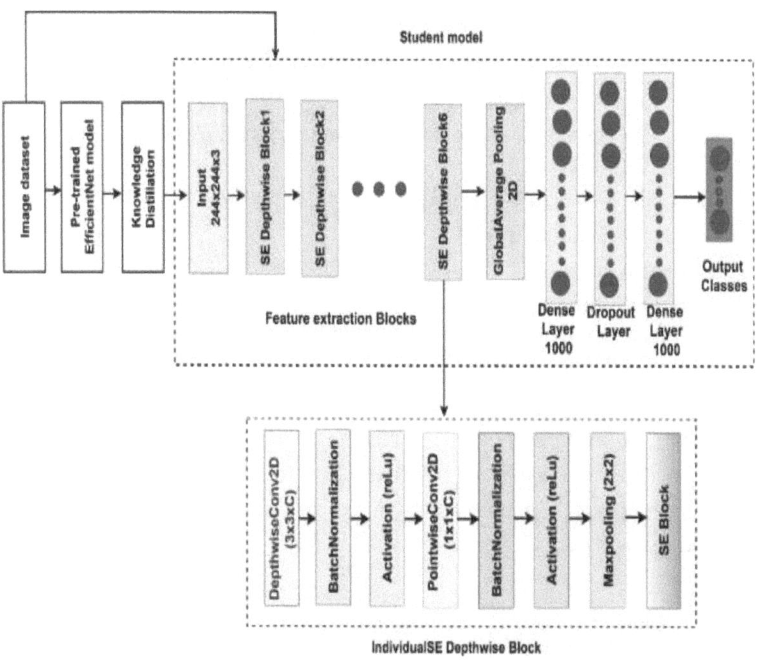

Tiny-LiteNet architecture

The pseudocode of the proposed AI-based IoT edge system appearing in Fig. 4 outlines the data processing chain between the stage of capturing the image and predicting the pest/disease, assessing confidence, and providing text-based reasons. The algorithm starts with the capturing of a live image stream through a camera and then through the lightweight model (Tiny-LiteNet) that runs on the Raspberry Pi 5, the images are then categorized into the applicable classes. The relevant interpretive text is then retrieved, thus enabling real-time, power-efficient and interpretable decision support of the agricultural monitoring in resource-constrained settings.

Fig. 4

Algorithm 1 AI-Powered IoT Edge System for Real-Time Pest and Disease Detection

```
Input: Real-time image stream from camera
Output: Pest/Disease class label, confidence score, and textual explanation
Initialize Tiny-LiteNet model
Load class labels and corresponding textual explanations
Initialize camera module and peripherals
while system is active do
    image ← camera.capture()
    image ← preprocess(image)                          ▷ Resize, normalize, reshape
    prediction ← TinyLiteNet.predict(image)
    predicted_class ← arg max(prediction)
    confidence ← max(prediction)
    explanation ← explanation_dict[predicted_class]
    Display or store: class label, confidence score, explanation
    if predicted_class is critical then
        Trigger alert
    end if
    Wait for next capture interval
end while
```

Pseudocode of the AI-powered IoT edge system with Tiny-LiteNet and text-based explainability

The performance in terms of training and validation of the Tiny-LiteNet model during the first 150 epochs is shown in Figure 5. Figure 5a shows that training and validation accuracies increase at a fast pace at the first ten epochs and then stabilize, reaching a validation rate of about 98.6', which testifies the excellent learning performance of the model. In Figure 5b, the training and validation losses decline sharply in the first twenty epochs, which is an additional sign of the effectiveness of the learning process of the model.

Fig. 5

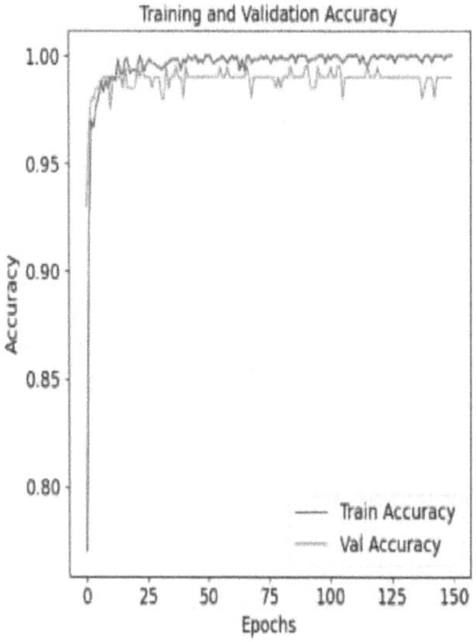

(a)

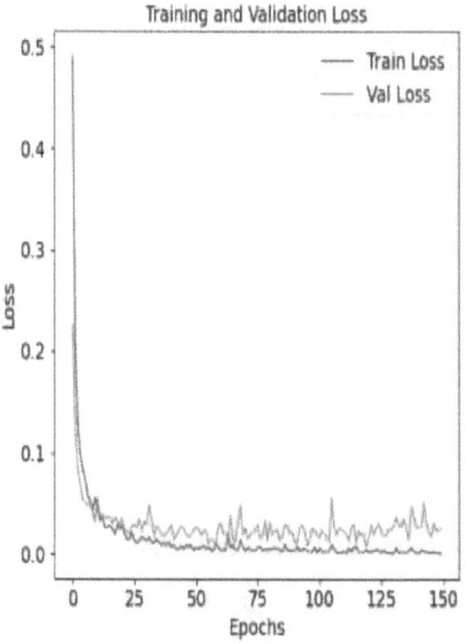

(b)

Figure (a) and (b) represent the performance accuracies and the performance losses of the model in the pest dataset with 150 epochs. As depicted in Fig. 6a, the training and validation accuracies of the model are high in the initial 20 epochs and the model achieves a validation accuracy of about 99 00 in and after that, which reflects the high-learning ability. Similarly, Fig. 6b indicates that both training and validation losses decrease fast during the initial 25 epochs indicating that the model learns successfully.

Fig. 6

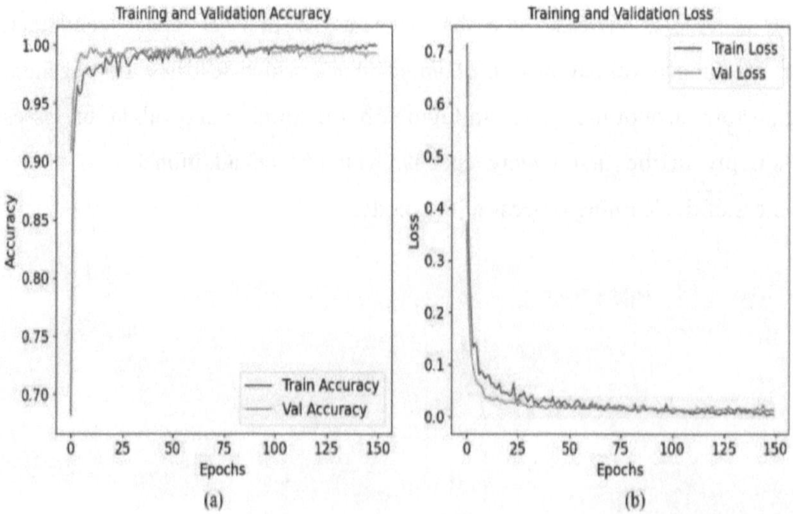

Plots (a) and (b) illustrate the performance accuracy and the loss values on the model on the disease dataset with 150 epochs. Figure 7 shows two confusion matrices, both of which describe how Tiny-LiteNet model performs when it predicts crop pests and diseases. Figure 7 a shows how the model classifies the diseases of the corn leaf, and Figure 7b shows its classification of crop pests. The model was exceptional in the performance, despite certain misunderstanding on Gray_leaf_spot disease and grasshopper pest.

Fig. 7

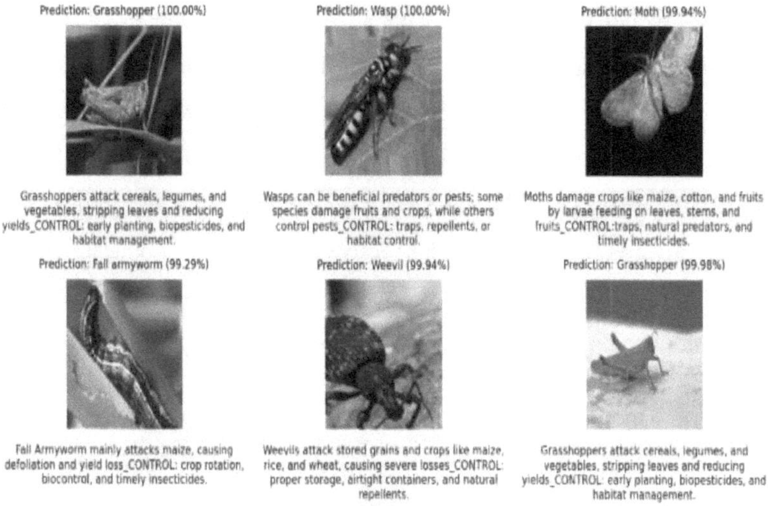

(a)

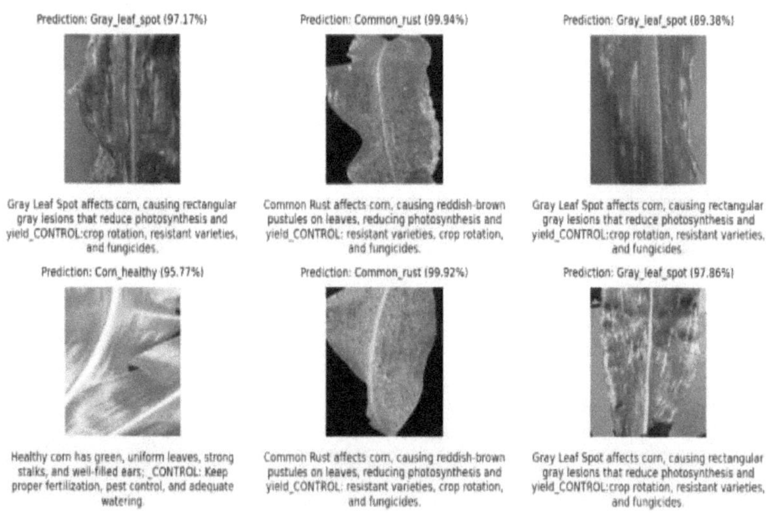

(b)

Confusion matrices for crop disease (**a**) and (**b**) crop pests

The Tiny-LiteNet model that is obtained herein, as seen in Figures 8a and 8b, forecasts and classifies images with a large precision level. The area deals with pictures of the health state of plants, such as maize crop pests, diseases, and fruit pests. In addition, the labelling of the prediction is highly accurate hence highlighting the effectiveness of the model in a wide variety of image-processing problems. Also, after classification, a descriptive summary is produced which summarizes the information about the image and gives control measures suggestions that could help in making informed decisions.

Fig. 8

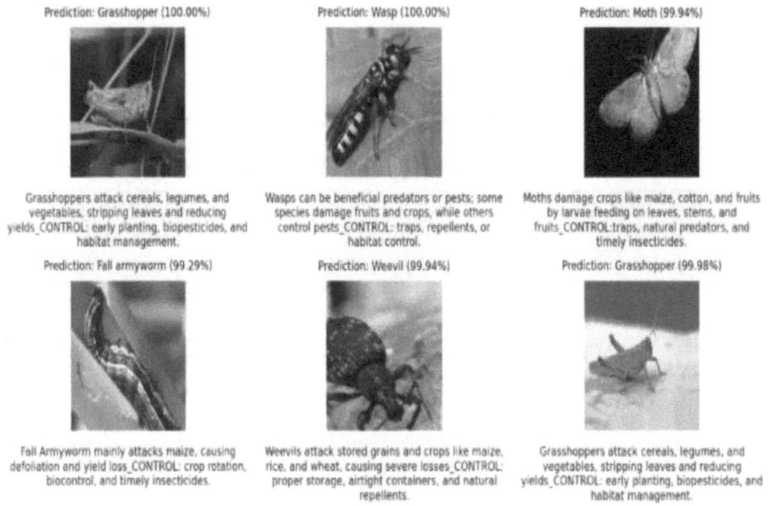

(a)

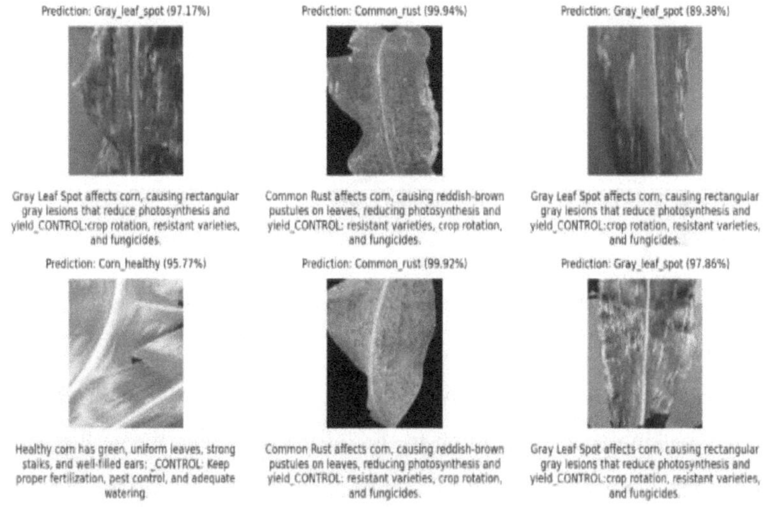

(b)

The table below gives predictions on (a) crop pest and (b) crop disease classification tasks and the description of the same.

Figure 9 shows a comparison between accuracy, storage efficiency, and responsiveness with eight convolutional neural network models, which underscores the effectiveness of Tiny-LiteNet to be used in resource-constrained deployments.

Fig. 9

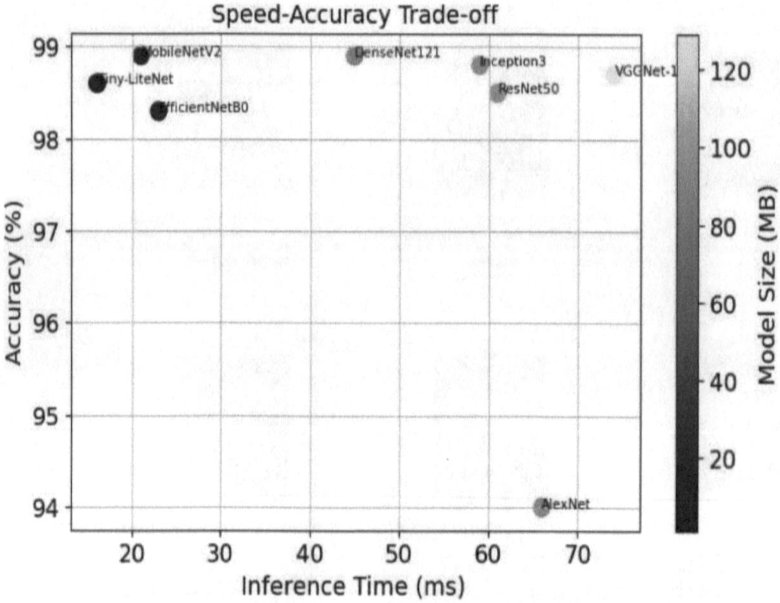

Accuracy and size of model visualisation and inference time of seven models. As shown in Figure 10, the portable edge device that would be used to detect crop pests and diseases has been executed and wrapped in a white cover. Fig. (a) shows the frontal view making special attention to the integrated screen where the user can interact and see the results. Figure (b) shows the back of it stating the camera lens used to take pictures and ports that can be opened to program and supply power, switch, and show the LED status on the top.

Fig. 10

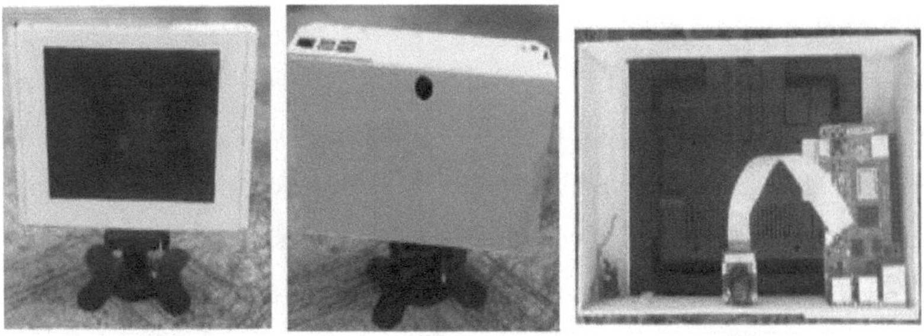

(a) (b) (c)

The bodily deployment of the IoT device is shown in Figure (a), which represents the frontside, Figure (b) which represents the backside, and Figure (c) which represents the internal elements. Combinations of the Tiny-LiteNet model and a Raspberry 5 are a powerful solution with high performance but lightweight and efficiency in detecting and classifying cereal crops pests and diseases. The proposed system also provides an explainability feature to the farmers to classify a pest or disease, the impact it has on plants, and the methods of controlling to reduce the risks. The model has very high accuracy, with a score of 98.6, an F1-score of 98.4, and a recall of 98.2, which is comparable or higher than most state-of-the-art deep-learning models, such as VGG-16, Inception-3, ResNet-50, DenseNet-121, EfficientNet-B0, and MobileNet-V2, which were trained on the same dataset. The differentiating feature of Tiny-LiteNet is its small size of 0.72MB, however, that allows it to be inferred quickly and with minimal computational cost and can run on edge-computing hardware and still has powerful predictive performance. The use of six squeeze-and-excitation (SE) blocks contributes to the feature extraction and the performance of the classification of the various pest and disease images. This design allows the model to differentiate between fine variations on pest and disease image features when conditions vary with respect to background, light variations and stage of growth and therefore enhances the generalization ability of the model. In addition, real time information transfer through GSM/GPRS can be used in remote monitoring and sending of messages to farmers on the status of pests and diseases. The capability of the system in edge-computing, portability and low power consumption make it affordable to small-holder farmers who have low resources. Machine Learning - Ideas in Crop Disease Detection. It is estimated that the world population will have reached a population that is larger than nine billion by the mid-century, thus creating a high demand in the food industry, and exerting more demands on the agricultural industry to yield higher production. Meeting this demand is complicated by a diverse range of issues, some of which are motivated by labour shortages, which are partially fuelled by rural-urban migration. At the same time, the sustainability of food production requires more responsibility due to the environmental problems like global warming, biodiversity loss, and soil degradation. Extreme weather conditions like droughts or floods also contribute to seasonality and increase the unpredictability of growing conditions as well as the likelihood of crop failures [56-62]. Besides, increasing temperatures and changes in the precipitation patterns enhance the development and geographic spread of pathogens of crops, and the decline in biodiversity decreases the natural physical barriers to the development of diseases, further increasing disease pressure in the field.[63, 64]. Pests and diseases are one of the most urgent challenges, and, altogether, they cause almost thirty percent of the crops losses worldwide, which is why it is necessary to develop efficient and timely disease detection and management methods. Traditional scanning and disease detection are more or less time-consuming, expensive, and subjective due to human intervention. In underprivileged areas, inaccessibility to agricultural specialists may result in even more loss of yields and make food insecurity worse [65, 66]. Precision agriculture has become increasingly popular as a data-driven system to respond to these threats to optimize inputs of farming activities, including fertilizers, pesticides, and water, to particular field conditions. Through the use of sensors, robotics, and smart computing, precision agriculture strives to maximize the efficiency and sustainability and minimize the impact on the environment. Abiotic and biotic stresses can be monitored and documented in real time using tools like drones and satellite platforms that are fitted with remote-sensing technology. However, multispectral or hyperspectral imaging is even more detailed in terms of information about plant health [67–69]. One of the central aspects of this high-technology change is artificial intelligence (AI), especially when it is being used with computer vision. Image analysis through AI allows detecting the symptoms of early-stage disease (i.e., leaf discoloration, fungal spots, etc.) automatically in large scale and at a significantly higher level than human perception (leaf discoloration, fungal spots, etc.) [70]. Such strategy can also be applied to reduce the use of pesticides since it allows precision and not blanket application of chemicals hence alleviating effects on the environment and human health [71]. Furthermore, better computational abilities and continuously growing amounts of data reinforce machine-learning and deep-learning models, which allow making more and more accurate diagnoses of plant pathologies. However, the quality of the data is another critical aspect of model performance; incorrect or un-representative data may be the cause of low generalizability.

These technical and environmental issues are accompanied by newly developed cybersecurity threats to digital agriculture systems. Cyber warfare can interfere with farm processes (hacking, manipulating data), or destroy the credibility of recommendations provided by AI. Likewise, agricultural IoT ransomware would interfere with real-time scanning data, which will compel farmers to make decisions without information. It is, thus, a significant aspect of the contemporary precision agriculture to assure a high level of data protection, secure communications channels, and modeling-integrity monitoring. The convergence of the necessity to have higher productivity, less chemical usage and robustness to climate instability makes the systematic review of AI-based methods of crop diseases recognition timely and essential. Combining the existing achievements of AI-based methods and analyzing real-life examples, scientists and industry experts will be less prone to misunderstandings about the areas in which the technologies perform admirably and where they are considerably inferior, particularly in low-resource areas or localities that experience extreme pest outbreaks. The subsequent paragraphs will briefly define the concept of precision agriculture and its technological basis and how the site-specific data collection and analysis approach could be employed to support informed decision-making in any farming setting. Then, a brief description of the most frequently used machine learning (ML) and computer vision (CV) algorithms to detect plant diseases is given, and their main strong points, possible weaknesses, and future security concerns are discussed equally. Thereafter, we present our methodology in conducting this systematic literature review, explaining our search, inclusion and exclusion criteria, data extraction, and the techniques of quality assessment applied to guarantee strong and free from bias results. In the results section, we provide a summary of our most important findings, showing how machine learning and deep learning-based disease detection model complexities have changed over the years in regard to the computational requirements, algorithm sophistication, and real-life examples of implementation. We emphasize such critical performance measures as accuracy, precision, recall, F1-score that were reported in the research, as well as information on model scalability, cost-efficiency, and environmental implications.

Subsequently, we discuss our section that gives meaning to the implications of these results in the wider context of precision agriculture. We also compare various AI-based solutions, highlight the best practices that have been discovered during the research, and also reveal the knowledge gaps in which future research can bring improvements. At last, we end by summarizing the main contributions of this review reiterating how AI-controlled disease detection holds a potential solution to key agricultural problems, and giving suggestions on direction of future research. Precision agriculture is a new method of agriculture which utilises new technologies to streamline farming operations, boost the efficiency of resources, and increase yields. The given methodology is focused on the field-specificity of the farming interventions and their adaptation to the unique conditions in individual fields or areas of fields so that the agricultural management is sustainable and highly efficient. Precision agriculture can be used to solve the increasing global food demand and reduce environmental degradation by utilising remote sensing, geographic information systems (GIS) and machine learning to manage food consumption. Precision agriculture relies on technological means that enable the gathering and analysis of information that supports the decision-making process. Remote sensing devices such as satellites, drones, and hyperspectral cameras are crucial in taking images of crops in fine details to observe the health and identify diseases. As an example, hyperspectral imaging has been used to monitor severity of diseases in crops so that farmers can deal with problems early enough. Moreover, the combination of artificial intelligence, specifically deep learning algorithms, has transformed the process of crop disease detection as the patterns and anomalies can be defined automatically in the images. Such algorithms include convolutional neural networks (CNNs), which are optimized using big datasets in order to improve their accuracy and flexibility to various farming contexts. One of the most important elements of the precision agriculture is site-specific data collection that provides interventions to be adapted to the specifics of each location within a field. Information is collected with the help of IoT sensors, UAVs (unmanned aerial vehicles), and other devices, which control soil moisture, temperature, and nutrient concentrations in real time. Indicatively, UAVs with multispectral cameras are commonly applied to locate stressed areas on crops that need prompt respondents. These tools enable fine mapping of variability of fields that are necessary to optimise outputs like fertilisers and pesticides. With these technologies, a lot of data would be generated and this data requires sophisticated methods of data analysis that can convert raw data into actionable information. Precision agriculture applies machine learning and big data analytics to analyze and understand multidimensional data. The approaches allow making predictions about disease outbreaks, resource optimisation, and yield estimation. As an example, machine-learning-driven predictive models have been trained to predict the health of crops and determine the effects of the environmental conditions that the farmers can use to direct their proactive management. With these tools of analysis, one is capable of a total overview of the crop performance and therefore resources are spent efficiently. The use of AI to identify crop disease is a fairly new but fast-changing phenomenon. The search result of eligible studies increased to 64 studies by 2024, and it increased to 64 compared to 1 article in 2016, which signifies a fast-paced innovation of AI-based crop-disease detection. Data is shown in Figure 11.

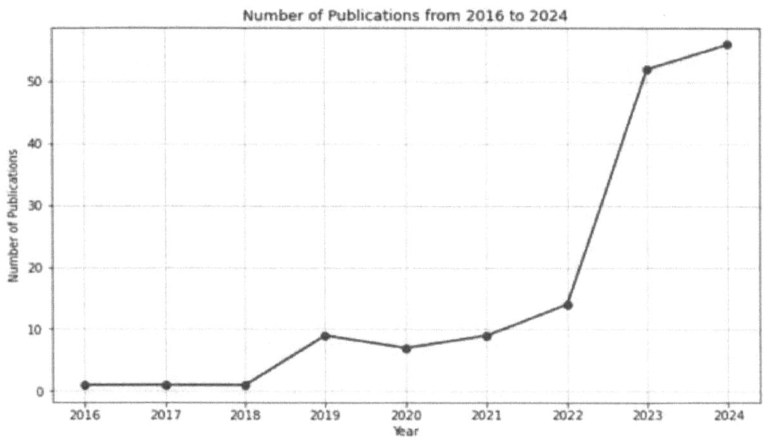

Figure 11. Expanding the number of publications between 2016 and 2024.

The amount of articles considered has allowed analyzing various AI models to detect disease in crops. SVM or KNN are classical machine-learning models that were used in 19 studies [81-99]. In 49 studies [100 -145], deep-learning models with the use of Convolutional Neural Networks (CNNs) were used. Also, the latest Vision Transformers (ViTs) were examined in 9 researches [146 148]. In addition, other authors chose to use hybrid AI methods, combining CNNs with standard ML solutions like SVM, which were found in 2 studies [153, 154]. This variety of AI models choice highlights the movement in the direction of using deep-learning models, in particular, CNNs and ViTs, as they outperform feature-extraction models. The popularity of CNNs and especially custom ones suggests optimisation of feature extraction towards domain-specific agricultural applications by researchers. Transfer learning is also a relevant method to use in utilizing pre-trained models with minimal computing needs. The combination of classical ML models with deep learning represents the attempts to achieve better generalisation, performance optimization, and model interpretability especially when small datasets are involved or agricultural information is structured. Our analysis results demonstrate that deep-, as compared to classical, machine-learning models are typically superior in terms of accuracy and precision. CNNs and Vision Transformers (ViTs) showed high classification results, and some of the studies reported above 99 percent accuracy [155-160]. Specifically, ViT-based models showed an impressive level of accuracy, and some of these studies reported scores approaching 100 percent [161-163]. There was also an excellent level of performance on CNN-based methods, with 99 or above being the usual performance, particularly with transfer learning optimisation [164-167]. Mean accuracy in these three main AI architectures is shown in Figure 12. On the contrary, the classical machine-learning models like SVM and KNN, although effective, generally showed lower levels of accuracy than deep-learning models did. This disparity is because they depend on hand-produced feature descriptors like colour histograms and texture measures, which both respond less to such fine-scale multi-scale patterns as are expressed by plant-disease images. In addition, most classical models were trained directly on somewhat small often imbalanced datasets, but CNN-, as well as ViT-, models are often trained on large-scale pre-training and large-scale data- augmentation pipelines, which contribute to improved generalisability to the variations in illumination, background, and crop species.

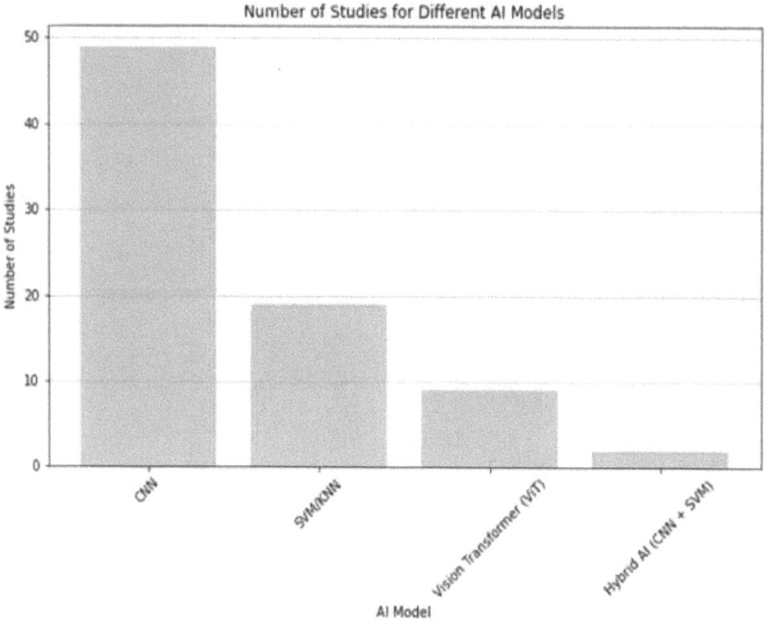

Figure 12. Number of studies for different AI models

Cost-effective and scalable information is not given in all the studies. Nevertheless, in the examples where such information is present [155-158], it can be observed that a large percentage of those utilise RGB images using publicly available data [134-137], thus making the data collection processes cheaper. Moreover, the majority of cost-efficient models are based on convolutional neural networks (CNNs) that have been trained using transfer learning which offers a tradeoff between performance and reduced computational demands. Therefore, despite the fact that Vision Transformer (ViT)-based models are more precise, CNN-based models are cheaper and more scalable [133-136] mainly because of their lower computational requirements and operation costs.

Figure 13,Figure 14

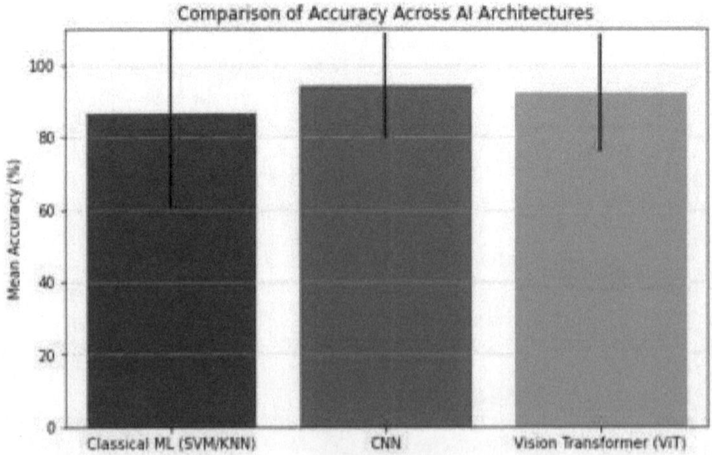

Figure 13. Comparison of accuracy across AI architectures.

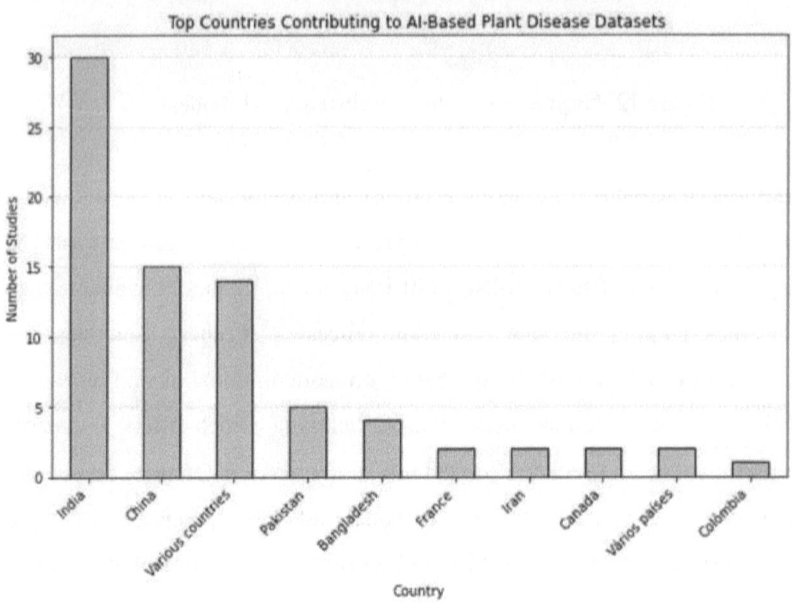

Figure 14. Top countries contributing to AI-based plant disease datasets.

Scalability can also be viewed as another major benefit of convolutional neural networks, and empirical data supports the idea of them being flexible in a variety of hardware settings as well as usability in real-time workloads [131, 132]. On the other hand, vision transformers, although more powerful, require very large amounts of computational power, which makes them more costly to implement and even less scalable in agricultural settings where the infrastructural constraints are a limiting factor [167, 168]. On this basis, vision transformers can be a better choice when the research is focused, and accuracy is the primary concern, whereas convolutional neural networks are more practical when the real-world applications take the center stage, and the elements of computational efficiency and cost are the primary factors. Figure 15 shows no significant dependence of dataset size on model performance, meaning that models that are trained on larger datasets do not necessarily have high accuracy.

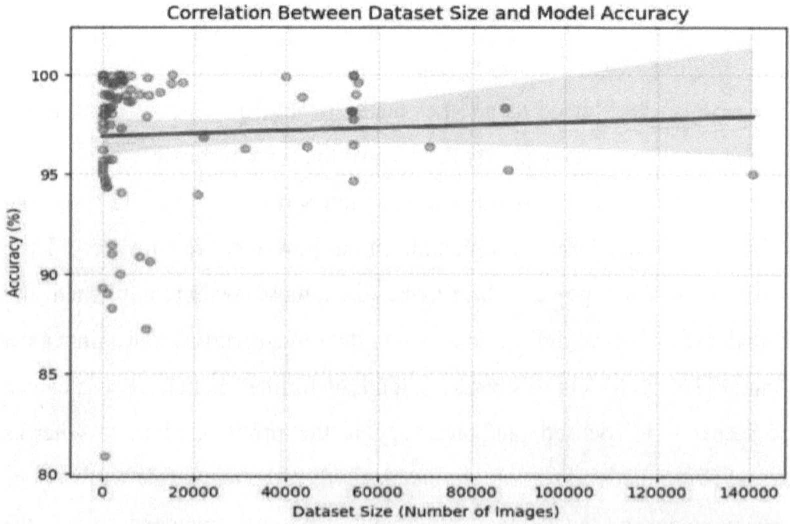

Figure 15. Correlation between dataset size and model accuracy

A summary is presented in Figure 16.

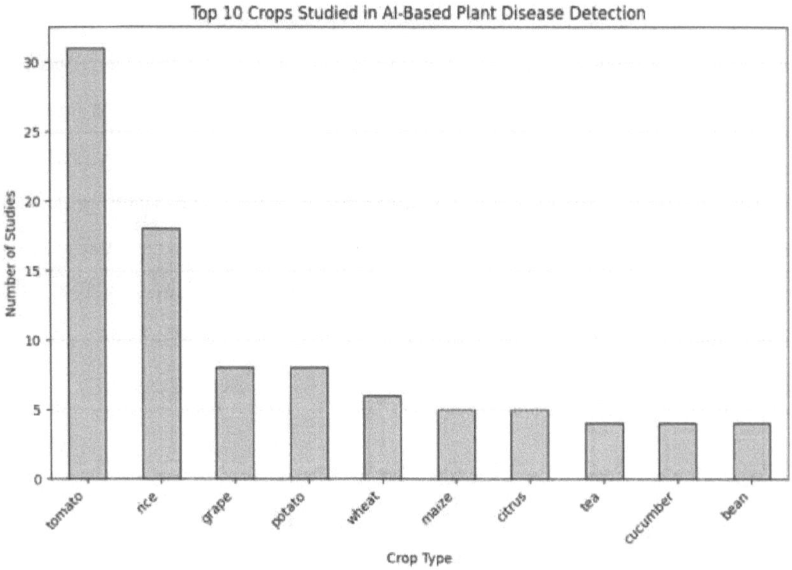

Figure 16. Top 10 crops in AI-based plant disease detection

The described trend might be explained by the social and economic importance of these staple crops that are critical to the world food security and serve as the main food resources of large groups of the population. Moreover, these crops are well-covered in open-source agricultural databases, especially in PlantVillage and other open-access repositories, as a result of which, AI models can be trained and benchmarked. This becomes more accessible as an incentive to more research in this field, unlike less commonly researched crops where there may not be available well-annnotated open repositories of images. Other crops including tea and grapes are also under extensive study due to their high economic value and impact in the agricultural markets of the world. These commodities are much sensitive to diseases, and therefore it is necessary to detect the disease early enough and avoid major economic losses due to poor yield. As an example, illnesses that affect tea and grapevines may lead to drastic reduction in production hence affecting international trades and local economies. In reference to the diseases to be studied, most of the research has been focused on fungal afflictions (97 studies), bacterial infections (65 studies), and viral diseases (20 studies). Fungal, bacterial, and viral are the three types of studies. However, despite more studies on fungal pathogens, the average percentage of detection of fungal disease detection (mean 88.85, SD 22.92) is lower than the interpretation of bacterial (mean 91.39, SD 18.69) or viral infections (mean 94.16, SD 10.68). Figure 17 is used to represent the results.

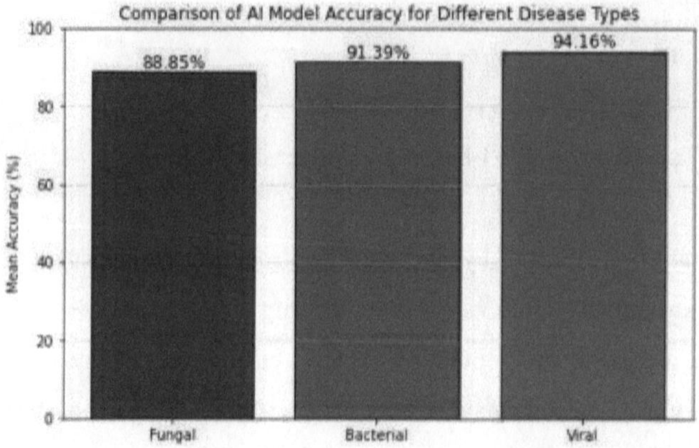

Figure 17. Comparison of AI model accuracy for difefrent disease types

The above section indicates some gaps in existing literature thus facilitating the suggested research directions and advancements in the future. The first gap is related to the geographic bias, because databases are disproportionately acquired in the Asian continent and not in other parts. Therefore, there is a definite demand of more diverse and representative data of underrepresented areas like Africa, Europe and South America. Increasing the diversity of the datasets will enhance the robustness, flexibility, and external validity of AI-based plant disease detection models, and thus make sure that the IT-based methodologies can be successfully applicable across different climates, crops, and other agricultural conditions. The processed data show that database size does not have a direct correlation with high model performance, especially in the area of accuracy. In addition, a large amount of research is based on images of publicly available RGB-databases, which are not always of high quality and might be limited to reflect the complexity of plant diseases. Thus, the diversity of the images, the quality of the images, and the accuracy of its annotations should be considered in future studies not only to find out the number of images or the size of databases but also to increase the strength of the AI models. To achieve high-quality images, it is recommended to take high-resolution images (at least 12 MP) of the image under the regulated lighting or portable light tents of the image at different viewing positions and developmental stages. Deep-learning-based models, including CNNs and Vision Transformers (ViT), which tend to perform better than conventional machine learning models like SVM and KNN were the most successful models. But deep learning models are the most computationally intensive and difficult to execute as well and can be scaled down. Moreover, there was a significant lack of hybrid models that combine deep learning with the conventional machine learning. In that aspect, it might be investigated in future studies to determine the hybrid methods in which the benefits of both techniques are used, striking some balance between reduced complexity of calculations and high precision. The majority of studies do not provide cost-effectiveness analysis, and the information regarding this area is limited. However, CNN-based models are relatively robust and capable of working with RGB image without using high-resolution input like multispectral or hyperspectral imagery. Hence, a low-cost model and infrastructures should be developed in the future to enable its implementation even in markets where the resources are limited. Additionally, these methods and models must be user-friendly, ideally, via cloud computing to reduce the costs of investment (e.g. the necessity to deploy high-performance computers and GPUs). It is also facilitated by the use of open-source software which further promotes cooperation, accessibility and research development in the field. The outcomes of this review show that deep learning models are very robust in the detection of crop diseases. It is important to note though that the researched models were so specific and had not been subjected to work in a great number of environments, including those with a great number of species, different climatic conditions or pest outbreaks. Transfer learning to adapt models that exist in other environments should also be examined further and more variety introduced in testing. Such a strategy may increase the flexibility of the models that have already been trained, which leads to lowering the financial and computational costs of training new models. There were also disparities in terms of the crops studied with the research concentrated on mostly rice, potatoes and tomatoes. Although this tendency can be partly attributed to their economic significance and contribution to the food security system, it is imperative to extend the scope of the research to other crops so that a more comprehensive and sustainable strategy towards the resilience of agriculture was developed. Moreover, fungi were given much more attention in the studies that were analyzed than bacterial or viral infections. Subsequent studies ought to strive to overcome this attention and focus especially on the research of less well-known diseases that could be a major economic and food security threat, especially in under-defended areas.

Finally, there is an urgent need to unify and analyze the dangers of technological solutions, in particular, the vulnerability to cyberattacks, both committed by malicious individuals and against state vehicles in time of war. Such technologies are especially susceptible to such dangers as data manipulation and information theft, deliberate shutdown of the system, and others. In this connection, the focus of the future studies should be on digital security, and it is important to focus on the creation of secure data transmission and storage methods, the preservation of the data integrity, and the protection against the external threats of the model.

Conclusion

The paper outlines the design and deployment of a lightweight, power efficient IoT edge architecture based on artificial intelligence, optimized to the real-time detection of crop pests and diseases in resource constrained agricultural settings. The experimental analysis shows that the proposed model has an overall accuracy of 98.6, F1 score of 98.4 and a recall of 98.2 with an inference latency of 80ms and compact model size of 1.2mb of 1.48 million parameters. These measures outperform a number of state-of-the-art CNN counterparts, including VGGNet-16, Inception -3, ResNet-50, DenseNet-121, EfficientNet-B0, and MobileNetV2. Additionally, the lightweight IoT -edge device operates data in situ, and thus does not require an internet connection, retains low power usage and is portable, thus making it applicable in resource limited environments. Currently, the system is run on a rule-based architecture and uses pre-programmed expertise knowledge to produce textual output as opposed to producing responses on the fly based on a language model since edge devices are computationally limited and have low memory. The next round of work will focus on increasing the ability of the model to generate automatic textual description of the health status of plants with the help of a large language model (LLM). And we also intend to dedicatedly limit power usage in order to support our future plans of living in the field with battery-powered options. Recently, the best crop disease detection model is a deep-learning model, especially convolutional neural networks (CNNs) and vision transformers (ViTs). The future progress in this area will be based on the development of various, well-labeled datasets and the establishment of lightweight systems that may be used in low-resource farm environments. To overcome geographic data gaps and commendable standards of annotation, it is necessary to translate the success of research into the reality. The next generation of AI research must focus on creating hybrid AI systems that will unite the performance of classical machine learning with the high accuracy of deep learning and balance the performance and the computational complexity. It is also important to focus more on the development of open and multi-regional datasets that would represent a large variety of crops, disease types, and environmental factors in order to increase model generalisability. Further research agenda involves how cost-effective and scalable solutions that could assist smallholder farmers in resource-constrained situations can be developed. Increasing AI pipelines in agriculture to cybersecurity measures to guarantee resistance against the adversarial attacks and manipulation of the data should also be a point of focus. By following these dimensions, the future work will speed up the process of transforming the use of AI to detect diseases in the experimental studies to the real and accessible tools in the agricultural industry.

References

1. Sylvre, N. & Jean, D. R. Updates on modern agricultural technologies adoption and its impacts on the improvement of agricultural activities in Rwanda: A review. *Int. J. Innov. Sci. Res. Technol.* **5**(12), 222–229 (2020).

2. Balogun, A. L., Adebisi, N., Abubakar, I. R., Dano, U. L. & Tella, A. Digitalization for transformative urbanization, climate change adaptation, and sustainable farming in Africa: Trend, opportunities, and challenges. *J. Integr. Environ. Sci.* **19**(1), 17–37. https://doi.org/10.1080/1943815X.2022.2033791 (2022).
3. Hanyurwimfura, D., Nizeyimana, E., Ndikumana, F., Mukanyiligira, D., Bakar Diwani, A. & Mukamanzi, F. Monitoring system to strive against fall armyworm in crops case study: Maize in Rwanda. In *Proc - 2018 IEEE SmartWorld, Ubiquitous Intell Comput Adv Trust Comput Scalable Comput Commun Cloud Big Data Comput Internet People Smart City Innov SmartWorld*/UIC/ATC/ScalCom/CBDCo. 2018;66–71.
4. Pierre, N. et al. AI based real-time weather condition prediction with optimized agricultural resources. *Eur. J. Technol.* 7(2), 36–49 (2023).
5. Taniushkina, D. et al. Case study on climate change effects and food security in Southeast Asia. *Sci. Rep.* **14**(1), 1–15. https://doi.org/10.1038/s41598-024-65140-y (2024).
6. Sahu, R. & Tripathi, P. A brief review on LPWAN technologies for large scale smart agriculture BT - advanced network technologies and intelligent computing. In: Verma A, Verma P, Pattanaik KK, Dhurandher SK, Woungang I, editors. p. 96–113 (Cham: Springer Nature Switzerland, 2024).
7. Kiobia, D. O. et al. A review of successes and impeding challenges of IoT-based insect pest detection systems for estimating agroecosystem health and productivity of cotton. *Sensors* **23**(8), 4127 (2023).
8. Rahman, W. et al. Automated detection of harmful insects in agriculture: A smart framework leveraging IoT, machine learning, and blockchain. *IEEE Trans. Artif. Intell.* **5**(9), 4787–4798 (2024).

9. Bueno, A. F. et al. Challenges for adoption of integrated pest management (IPM): The soybean example. *Neotrop. Entomol.* **50**(1), 5–20. https://doi.org/10.1007/s13744-020-00792-9 (2021).
10. Shahbazi, F., Shahbazi, S., Nadimi, M. & Paliwal, J. Losses in agricultural produce: A review of causes and solutions, with a specific focus on grain crops. *J. Stored Prod. Res.* **111**, 102547 (2025).
11. Pascoal, D. et al. A technical survey on practical applications and guidelines for IoT sensors in precision agriculture and viticulture. *Sci. Rep.* **14**(1), 90404 (2024).
12. Delfani, P., Thuraga, V., Banerjee, B. & Chawade, A. Integrative approaches in modern agriculture: IoT, ML and AI for disease forecasting amidst climate change. *Precis Agric.* **25**, 2589–2613 (2024).
13. Morchid, A. et al. Integrated internet of things (IoT) solutions for early fire detection in smart agriculture. *Results Eng.* **24**(September), 103392. https://doi.org/10.1016/j.rineng.2024.103392 (2024).
14. Hakorimana, F. & Akçaöz, H. The climate change and Rwandan coffee sector. *Turkish J. Agric. - Food Sci. Technol.* **5**(10), 1206 (2017).
15. Saini, A. et al. Smart crop disease monitoring system in IoT using optimization enabled deep residual network. *Sci. Rep.* **15**(1), 1456 (2025).
16. de la San Emeterio, P. M., Martínez-Ortega, J. F., Castillejo, P. & Lucas-Martínez, N. Spatio-temporal semantic data management systems for IoT in agriculture 50: Challenges and future directions. *Internet of Things (Netherlands)* **2024**(25), 101030. https://doi.org/10.1016/j.iot.2023.101030 (2023).

17. Zhang, X., Cao, Z. & Dong, W. Overview of edge computing in the agricultural internet of things: Key technologies, applications, challenges. *IEEE Access* **8**, 141748–141761 (2020).
18. Elumalai, M., Fernandez, T. F. & Ragab, M. Machine Learning (ML) Algorithms on IoT and drone data for smart farming BT - intelligent robots and drones for precision agriculture. In Balasubramanian S, Natarajan G, Chelliah PR, editors. p. 179–206 (Cham: Springer Nature Switzerland; 2024). https://doi.org/10.1007/978-3-031-51195-0_10
19. Bao, W., Fan, T., Hu, G., Liang, D. & Li, H. Detection and identification of tea leaf diseases based on AX-RetinaNet. *Sci. Rep.* **12**(1), 1–16. https://doi.org/10.1038/s41598-022-06181-z (2022).
20. Kumbhar, V., Patil, A., Kumari, S. & Bharti, N. Systematic review on growth of E-agriculture in context of android-based mobile applications BT - ICT systems and sustainability. In: Tuba M, Akashe S, Joshi A, editors. p. 545–53 (Singapore: Springer Nature Singapore; 2023).
21. Khan, S. et al. Comparative analysis of deep neural network architectures for renewable energy forecasting: Enhancing accuracy with meteorological and time-based features. *Discov. Sustain.* **5**(1), 1–24. https://doi.org/10.1007/s43621-024-00783-5 (2024).
22. Tang, F. H. M., Lenzen, M., McBratney, A. & Maggi, F. Risk of pesticide pollution at the global scale. *Nat. Geosci.* **14**(4), 206–210. https://doi.org/10.1038/s41561-021-00712-5 (2021).
23. Khan, S. et al. Future of sustainable farming: exploring opportunities and overcoming barriers in drone-IoT integration. *Discov. Sustain.* **5**(1), 1–22. https://doi.org/10.1007/s43621-024-00736-y (2024).

24. Saleem, M. H., Potgieter, J., Arif, K. M. Automation in agriculture by machine and deep learning techniques: A review of recent developments. Vol. 22, 2053–2091 p. (Precision Agriculture. Springer US; 2021). https://doi.org/10.1007/s11119-021-09806-x
25. Kaplun, D. et al. An intelligent agriculture management system for rainfall prediction and fruit health monitoring. *Sci. Rep.* **14**(1), 1–23. https://doi.org/10.1038/s41598-023-49186-y (2024).
26. Garg, G. et al. CROPCARE: An intelligent real-time sustainable IoT System for crop disease detection using mobile vision. *IEEE Internet Things J.* **10**(4), 2840–2851 (2023).
27. Nayagam, M. G. et al. Control of pests and diseases in plants using IOT Technology. *Meas Sens.* **26**(December 2022), 100713. https://doi.org/10.1016/j.measen.2023.100713 (2023).
28. Ouhami, M., Hafiane, A., Es-Saady, Y., El Hajji, M. & Canals, R. Computer vision, IoT and data fusion for crop disease detection using machine learning: A survey and ongoing research. *Remote Sens.* **13**(13), 2486 (2021).
29. Azfar, S. et al. IoT-based cotton plant pest detection and smart-response system. *Appl. Sci.* **13**(3), 1851 (2023).
30. Anam, I. et al. Smart Agricultural Technology A systematic review of UAV and AI integration for targeted disease detection, weed management, and pest control in precision agriculture. *Smart Agric. Technol.* **9**(November), 100647. https://doi.org/10.1016/j.atech.2024.100647 (2024).
31. Vijayan, S. & Chowdhary, C. L. Hybrid feature optimized CNN for rice crop disease prediction. *Sci. Rep.* **15**(1), 1–20 (2025).

32. Ullah, N. et al. An efficient approach for crops pests recognition and classification based on novel DeepPestNet deep learning model. *IEEE Access* **10**(May), 73019–73032 (2022).

33. Albattah, W., Javed, A., Nawaz, M., Masood, M. & Albahli, S. Artificial intelligence-based drone system for multiclass plant disease detection using an improved efficient convolutional neural network. *Front. Plant Sci.* **13**(June), 1–16 (2022).

34. Karar, M. E., Alsunaydi, F., Albusaymi, S. & Alotaibi, S. A new mobile application of agricultural pests recognition using deep learning in cloud computing system. *Alexandria Eng. J.* **60**(5), 4423–4432. https://doi.org/10.1016/j.aej.2021.03.009 (2021).

35. Nyakuri, J. P., Nkundineza, C., Gatera, O. & Nkurikiyeyezu, K. State-of-the-art deep learning algorithms for internet of things-based detection of crop pests and diseases: A comprehensive review. *IEEE Access* **12**(August), 169824–169849 (2024).

36. Khan, A. et al. AI-enabled crop management framework for pest detection using visual sensor data. *Plants* **13**(5), 653 (2024).

37. Fu, R. et al. Machine-learning-based UAV-assisted agricultural information security architecture and intrusion detection. *IEEE Internet Things J.* **10**(21), 18589–18598 (2023).

38. Naseer, A. et al. A systematic literature review of the IoT in agriculture—Global adoption, innovations, security, and privacy challenges. *IEEE Access* **12**(May), 60986–61021 (2024).

39. Sinha, P. et al. OPEN A high performance hybrid LSTM CNN secure architecture for IoT environments using deep learning. *Sci. Rep.* **15**, 9684 (2025).

40. Ibrahim, A. S. et al. AI-IoT based smart agriculture pivot for plant diseases detection and treatment. *Sci. Rep.* **15**(1), 1–16 (2025).

41. Ahmed, S. et al. IoT based intelligent pest management system for precision agriculture. *Sci. Rep.* **14**(1), 1–15 (2024).

42. Nguyen-Tan, T. & Le-Trung, Q. A Novel 5G PMN-driven approach for AI-powered irrigation and crop health monitoring. *IEEE Access* **12**(September), 125211–125222 (2024).

43. Toscano, F. et al. Unmanned aerial vehicle for precision agriculture: A review. *IEEE Access* **12**(May), 69188–69205 (2024).

44. van Hilten, M. & Wolfert, S. 5G in agri-food - A review on current status, opportunities and challenges. *Comput. Electron. Agric.* **201**(March), 107291. https://doi.org/10.1016/j.compag.2022.107291 (2022).

45. El, M. A., Tatane, K. & Chihab, Y. Transforming agricultural supply chains: Leveraging blockchain-enabled java smart contracts and IoT integration. *ICT Express* **10**(3), 650–672. https://doi.org/10.1016/j.icte.2024.03.007 (2024).

46. Hasan, H. R. et al. Smart agriculture assurance: IoT and blockchain for trusted sustainable produce. *Comput. Electron. Agric.* **224**(February), 109184. https://doi.org/10.1016/j.compag.2024.109184 (2024).

47. Raja, S. R., Subashini, B., Prabu, R. S. 5G Technology in smart farming and its applications BT - intelligent robots and drones for precision agriculture. In: Balasubramanian S, Natarajan G, Chelliah PR, editors. p. 241–64 (Cham: Springer Nature Switzerland; 2024). https://doi.org/10.1007/978-3-031-51195-0_12

48. Praveen Kumar, M., Abhishek, A. B., Pavanalaxmi, S., Kripa, T. & Nayak, R. Emerging trends and future directions in ML for drone-assisted IoT networks BT - machine learning for drone-enabled IoT networks: Opportunities,

developments, and trends. In: Hassan J, Khalifa S, Misra P, editors. p. 191–207 (Cham: Springer Nature Switzerland; 2025). https://doi.org/10.1007/978-3-031-80961-3_10

49. Ghadi, Y. Y. et al. Integration of federated learning with IoT for smart cities applications, challenges, and solutions. *PeerJ Comput. Sci.* **9**, 1–23 (2023).
50. Mrabet, H., Belguith, S., Alhomoud, A. & Jemai, A. A survey of IoT security based on a layered architecture of sensing and data analysis. *Sensors (Switzerland)* **20**(13), 1–20 (2020).
51. Zhang, J. et al. ISMSFuse: Multi-modal fusing recognition algorithm for rice bacterial blight disease adaptable in edge computing scenarios. *Comput. Electron. Agric.* **223**(April), 109089. https://doi.org/10.1016/j.compag.2024.109089 (2024).
52. Gookyi, D. A. N. et al. TinyML for smart agriculture: Comparative analysis of TinyML platforms and practical deployment for maize leaf disease identification. *Smart Agric. Technol.* **8**(April), 100490. https://doi.org/10.1016/j.atech.2024.100490 (2024).
53. Karim, M. J. et al. Enhancing agriculture through real-time grape leaf disease classification via an edge device with a lightweight CNN architecture and Grad-CAM. *Sci. Rep.* **14**(1), 1–23. https://doi.org/10.1038/s41598-024-66989-9 (2024).
54. Zhang, F. et al. LSANNet: A lightweight convolutional neural network for maize leaf disease identification. *Biosyst. Eng.* **248**, 97–107 (2024).
55. Zeng, M., Chen, S., Liu, H., Wang, W. & Xie, J. HCFormer: A lightweight pest detection model combining CNN and ViT. *Agronomy* **14**(9), 1940 (2024).

56. United Nations. World Population Prospects 2022: Summary of Results. Erscheinungsort nicht ermittelbar: United Nations; 2022.
57. Food and Agriculture Organization of the United Nations (FAO). How to Feed the World in 2050. 2009.
58. D. Ge, H. Long, W. Qiao, Z. Wang, D. Sun, R. YangEffects of rural–urban migration on agricultural transformation: A case of Yucheng City, ChinaJournal of Rural Studies., 76 (2020), pp. 85-95
59. A.E. Hill, I. Ornelas, JE. Taylor Agricultural Labor Supply, Annu Rev Resour Econ, 13 (2021), pp. 39-64
60. Y. Yang, D. Tilman, Z. Jin, P. Smith, C.B. Barrett, Y-G Zhu, *et al.* Climate change exacerbates the environmental impacts of agriculture, Science, 385 (2024), p. eadn3747
61. Y. Kunimitsu, G. Sakurai, T. Iizumi Systemic Risk in Global Agricultural Markets and Trade Liberalization under Climate Change: Synchronized Crop-Yield Change and Agricultural Price Volatility, Sustainability, 12 (2020), Article 10680
62. T. Hasegawa, G. Sakurai, S. Fujimori, K. Takahashi, Y. Hijioka, T. Masui Extreme climate events increase risk of global food insecurity and adaptation needs, Nat Food, 2 (2021), pp. 587-595
63. A. PfenningButterworth, L.B. Buckley, J.M. Drake, J.E. Farner, M.J. Farrell, A-LM Gehman, *et al.* Interconnecting global threats: climate change, biodiversity loss, and infectious diseases, The Lancet Planetary Health, 8 (2024), pp. e270-e283
64. D. Kumar, R. Mukhopadhyay Climate change and plant pathogens: Understanding dynamics, risks and mitigation strategies, Plant Pathology, 74 (2025), pp. 59-68

65. M. Junaid, A. Gokce GLOBAL AGRICULTURAL LOSSES AND THEIR CAUSES, Bull Biol All Sci Res, 2024 (2024), p. 66
66. A. Ahmad, D. Saraswat, A. El Gamal A survey on using deep learning techniques for plant disease diagnosis and recommendations for development of appropriate tools, Smart Agricultural Technology, 3 (2023), Article 100083
67. G. Mgendi Unlocking the potential of precision agriculture for sustainable farming, Discov Agric, 2 (2024), p. 87
68. S.S. Dhaliwal, V. Sharma, Y.S. Shivay, R.K. Gupta, V. Verma, M. Kaur, *et al.* Assessment and detection of biotic and abiotic stresses in field crops through remote and proximal sensing techniques—evidence from earlier findings, Arab J Geosci, 17 (2024), p. 188
69. R. Rayhana, Z. Ma, Z. Liu, G. Xiao, Y. Ruan, J.S. Sangha A Review on Plant Disease Detection Using Hyperspectral Imaging, IEEE Trans Agri Elect, 1 (2023), pp. 108-134
70. U. Khan, M.K. Khan, M.A. Latif, M. Naveed, M.M. Alam, S.A. Khan, *et al.* A Systematic Literature Review of Machine Learning and Deep Learning Approaches for Spectral Image Classification in Agricultural Applications Using Aerial Photography, CMC, 78 (2024), pp. 2967-3000
71. S. Getahun, H. Kefale, Y. Gelaye Application of Precision Agriculture Technologies for Sustainable Crop Production and Environmental Sustainability: A Systematic Review, The Scientific World Journal, 2024 (2024), Article 2126734
72. A.N. Alahmadi, S.U. Rehman, H.S. Alhazmi, D.G. Glynn, H. Shoaib, P. Solé Cyber-Security Threats and Side-Channel Attacks for Digital Agriculture, Sensors, 22 (2022), p. 3520

73. M. Campoverde-Molina, S. Luján-Mora Cybersecurity in smart agriculture: A systematic literature review, Computers & Security, 150 (2025), Article 104284

74. Y. Gao, S.A. Camtepe, N.H. Sultan, H.T. Bui, A. Mahboubi, H. Aboutorab, *et al.* Security threats to agricultural artificial intelligence: Position and perspective, Computers and Electronics in Agriculture, 227 (2024), Article 109557

75. Z. Angyalos, S. Botos, R. Szilagyi The importance of cybersecurity in modern agriculture, JAI, 12 (2022)

76. L.C. Ngugi, M. Abelwahab, M. Abo-Zahhad Recent advances in image processing techniques for automated leaf pest and disease recognition – A review, Information Processing in Agriculture, 8 (2021), pp. 27-51

77. M. Kganyago, C. Adjorlolo, P. Mhangara, L. Tsoeleng Optical remote sensing of crop biophysical and biochemical parameters: An overview of advances in sensor technologies and machine learning algorithms for precision agriculture, Computers and Electronics in Agriculture, 218 (2024), Article 108730

78. A. Rejeb, K. Rejeb, S. Zailani, J.G. Keogh, A. Appolloni Examining the interplay between artificial intelligence and the agri-food industry, Artificial Intelligence in Agriculture, 6 (2022), pp. 111-128

79. M. Ariza-Sentís, S. Vélez, R. Martínez-Peña, H. Baja, J. Valente Object detection and tracking in Precision Farming: a systematic review, Computers and Electronics in Agriculture, 219 (2024), Article 108757

80. V.G. Dhanya, A. Subeesh, N.L. Kushwaha, D.K. Vishwakarma, T. Nagesh Kumar, G. Ritika, *et al.* Deep learning based computer vision approaches for

smart agricultural applications Artificial Intelligence in Agriculture, 6 (2022), pp. 211-229

81. P. Varalakshmi, S. Aravindkumar Plant disorder precognition by image based pattern recognition Procedia Computer Science, 165 (2019), pp. 502-510

82. Y. Wang, U.S. RajkumarDhamodharan, N. Sarwar, F.A. Almalki, Q.H. Naith, S. R, *et al.* A hybrid approach for rice crop disease detection in agricultural IoT system, Discov Sustain, 5 (2024), p. 99

83. A. Chug, A. Bhatia, A.P. Singh, D. Singh A novel framework for image-based plant disease detection using hybrid deep learning approach, Soft Comput, 27 (2023), pp. 13613-13638

84. A.N. S, H. R, R. S, S. Gaur, S. R A smart agriculturing IoT system for banana plants disease detection through inbuilt compressed sensing devices, Multimed Tools Appl, 82 (2023), pp. 45519-45539

85. S. Saleki, J. Tahmoresnezhad Agry: a comprehensive framework for plant diseases classification via pretrained EfficientNet and convolutional neural networks for precision agriculture, Multimed Tools Appl, 83 (2024), pp. 64813-64851

86. X.E. Pantazi, D. Moshou, AA. Tamouridou Automated leaf disease detection in different crop species through image features analysis and One Class Classifiers, Computers and Electronics in Agriculture, 156 (2019), pp. 96-104

87. R. Karthickmanoj, T. Sasilatha, J. Padmapriya Automated machine learning based plant stress detection system Materials Today: Proceedings, 47 (2021), pp. 1887-1891

88. S. Zhang, W. Huang, H. Wang Crop disease monitoring and recognizing system by soft computing and image processing models, Multimed Tools Appl, 79 (2020), pp. 30905-30916

89. A. Bhola, P. Kumar Deep feature-support vector machine based hybrid model for multi-crop leaf disease identification in Corn, Rice, and Wheat, Multimed Tools Appl (2024), 10.1007/s11042-024-18733-8
90. K. R, S. T Development of plant disease detection for smart agriculture, Multimed Tools Appl, 83 (2023), pp. 54391-54410
91. A. Adeel, M.A. Khan, M. Sharif, F. Azam, J.H. Shah, T. Umer, *et al.* Diagnosis and recognition of grape leaf diseases: An automated system based on a novel saliency approach and canonical correlation analysis based multiple features fusion, Sustainable Computing: Informatics and Systems, 24 (2019), Article 100349
92. G. Mustafa, H. Zheng, I.H. Khan, J. Zhu, T. Yang, A. Wang, *et al.*Enhancing fusarium head blight detection in wheat crops using hyperspectral indices and machine learning classifiers, Computers and Electronics in Agriculture, 218 (2024), Article 108663
93. A. Abdalla, H. Cen, L. Wan, R. Rashid, H. Weng, W. Zhou, *et al.* Fine-tuning convolutional neural network with transfer learning for semantic segmentation of ground-level oilseed rape images in a field with high weed pressure, Computers and Electronics in Agriculture, 167 (2019), Article 105091
94. H. Zhu, B. Chu, C. Zhang, F. Liu, L. Jiang, Y. He Hyperspectral Imaging for Presymptomatic Detection of Tobacco Disease with Successive Projections Algorithm and Machine-learning, Classifiers. Sci Rep., 7 (2017), p. 4125
95. S. Nurul Afiah Mohd Johari, S. Khairunniza-Bejo, A. Rashid Mohamed Shariff, N. Azuan Husin, M. Mazmira Mohd Basri, N. Kamarudin Identification of bagworm (Metisa plana) instar stages using hyperspectral

imaging and machine learning techniques, Computers and Electronics in Agriculture, 194 (2022), Article 106739

96. J.B. Motie, M.H. Saeidirad, M. Jafarian Identification of Sunn-pest affected (Eurygaster Integriceps put.) wheat plants and their distribution in wheat fields using aerial imaging, Ecological Informatics, 76 (2023), Article 102146
97. B.V. Nikith, N.K.S. Keerthan, M.S. Praneeth, Amrita DrT Leaf Disease Detection and Classification, Procedia Computer Science, 218 (2023), pp. 291-300
98. Mia MdR, S. Roy, S.K. Das, Rahman MdA Mango leaf disease recognition using neural network and support vector machine Iran J Comput Sci, 3 (2020), pp. 185-193
99. R. Sujatha, J.M. Chatterjee, N. Jhanjhi, SN. Brohi Performance of deep learning vs machine learning in plant leaf disease detection Microprocessors and Microsystems, 80 (2021), Article 103615
100. A. Usha Ruby, J. George Chellin Chandran, B.N. Chaithanya, T.J. Swasthika Jain, R. Patil Wheat leaf disease classification using modified ResNet50 convolutional neural network model, Multimed Tools Appl, 83 (2024), pp. 62875-62893
101. A. Karlekar, A. Seal SoyNet: Soybean leaf diseases classification Computers and Electronics in Agriculture, 172 (2020), Article 105342
102. G. Heller, E. Perrin, V. Vrabie, C. Dusart, ML Panon, M. Loyaux, *et al.* Multisource neural network feature map fusion: An efficient strategy to detect plant diseases, Intelligent Systems with Applications, 19 (2023), Article 200264
103. K. Perveen, S. Debnath, B. Pandey, S.P. Chand, N.A. Bukhari, P. Bhowmick, *et al.* Deep learning-based multiscale CNN-based U network model for leaf

disease diagnosis and segmentation of lesions in tomato, Physiological and Molecular Plant Pathology, 128 (2023), Article 102148

104. V. Rajpoot, A. Tiwari, AS. Jalal Automatic early detection of rice leaf diseases using hybrid deep learning and machine learning methods, Multimed Tools Appl, 82 (2023), pp. 36091-36117

105. N.G. Rezk, A-F Attia, M.A. El-Rashidy, A. El-Sayed, EE-D. Hemdan An Efficient Plant Disease Recognition System Using Hybrid Convolutional Neural Networks (CNNs) and Conditional Random Fields (CRFs) for Smart IoT Applications in Agriculture, Int J Comput Intell Syst, 15 (2022), p. 65

106. T. Wang, H. Xu, Y. Hai, Y. Cui, Z. Chen An Improved Crop Disease Identification Method Based on Lightweight Convolutional Neural Network, Journal of Electrical and Computer Engineering, 2022 (2022), pp. 1-16

107. S. Anwarul, M. Mohan, R. Agarwal An Unprecedented Approach for Deep Learning Assisted Web Application to Diagnose Plant Disease, Procedia Computer Science, 218 (2023), pp. 1444-1453

108. S. Ashwinkumar, S. Rajagopal, V. Manimaran, B. Jegajothi Automated plant leaf disease detection and classification using optimal MobileNet based convolutional neural networks, Materials Today: Proceedings, 51 (2022), pp. 480-487

109. M.M. Ozguven, K. Adem Automatic detection and classification of leaf spot disease in sugar beet using deep learning algorithms, Physica A: Statistical Mechanics and its Applications, 535 (2019), Article 122537

110. J. Gao, J.C. Westergaard, E.H.R. Sundmark, M. Bagge, E. Liljeroth, E. Alexa ndersson Automatic late blight lesion recognition and severity quantification based on field imagery of diverse potato genotypes by deep learning, Knowledge-Based Systems, 214 (2021), Article 106723

111. S. Talasila, K. Rawal, G. Sethi Black gram disease classification using a novel deep convolutional neural network, Multimed Tools Appl, 82 (2023), pp. 44309-44333
112. A. Kumar, VK. Patel Classification and identification of disease in potato leaf using hierarchical based deep learning convolutional neural network, Multimed Tools Appl, 82 (2023), pp. 31101-31127
113. M.G. Lanjewar, JS. Parab CNN and transfer learning methods with augmentation for citrus leaf diseases detection using PaaS cloud on mobile, Multimed Tools Appl, 83 (2023), pp. 31733-31758
114. M.G. Lanjewar, KG. Panchbhai Convolutional neural network based tea leaf disease prediction system on smart phone using paas cloud, Neural Comput & Applic, 35 (2023), pp. 2755-2771
115. S. Zhang, S. Zhang, C. Zhang, X. Wang, Y. Shi Cucumber leaf disease identification with global pooling dilated convolutional neural network, Computers and Electronics in Agriculture, 162 (2019), pp. 422-430
116. S. Mishra, R. Sachan, D. Rajpal Deep Convolutional Neural Network based Detection System for Real-time Corn Plant Disease Recognition, Procedia Computer Science, 167 (2020), pp. 2003-2010
117. A. Haridasan, J. Thomas, E.D. Raj Deep learning system for paddy plant disease detection and classification, Environ Monit Assess, 195 (2023), p. 120
118. E. Elfatimi, R. Eryiğit, L. Elfatimi Deep multi-scale convolutional neural networks for automated classification of multi-class leaf diseases in tomatoes, Neural Comput & Applic, 36 (2024), pp. 803-822
119. S. Sladojevic, M. Arsenovic, A. Anderla, D. Culibrk, D. Stefanovi Deep Neural Networks Based Recognition of Plant Diseases by Leaf Image

Classification, Computational Intelligence and Neuroscience, 2016 (2016), pp. 1-11

120. V. Sharma, S. Kannan, S. Tanya, N. Panda Detecting Plant Diseases at Scale: A Distributed CNN Approach with PySpark and Hadoop, Procedia Computer Science, 235 (2024), pp. 1044-1057

121. M. Sarah, M. Abdlemadjid, B. Sarah, H. Yacine, CEM. Miloud Evaluating the effect of super-resolution for automatic plant disease detection: application to potato late blight detection, Multimed Tools Appl., 83 (2024), pp. 78469-78487

122. T. Wang, X. Mei, J. AlexThomasson, C. Yang, X. Han, P.K. Yadav, *et al.* GIS-based volunteer cotton habitat prediction and plant-level detection with UAV remote sensing, Computers and Electronics in Agriculture, 193 (2022), Article 106629

123. G. G, A.P. J Identification of plant leaf diseases using a nine-layer deep convolutional neural network, Computers & Electrical Engineering, 76 (2019), pp. 323-338

124. D. Mamba Kabala, A. Hafiane, L. Bobelin, R. Canals Image-based crop disease detection with federated learning, Sci Rep, 13 (2023), Article 19220

125. R. Hadipour-Rokni, E. Askari Asli-Ardeh, A. Jahanbakhshi, I. Esmaili paeen-Afrakoti, S. Sabzi Intelligent detection of citrus fruit pests using machine vision system and convolutional neural network through transfer learning technique, Computers in Biology and Medicine, 155 (2023), Article 106611

126. S. Raman, M. Soni, R. Ramaprasad, V. Chamola LWCNN: a lightweight convolutional neural network for agricultural crop protection, Multimed Tools Appl, 81 (2022), pp. 22323-22334

127. S. Mishra, D.R. Volety, N. Bohra, S. Alfarhood, M. Safran A smart and sustainable framework for millet crop monitoring equipped with disease detection using enhanced predictive intelligence, Alexandria Engineering Journal, 83 (2023), pp. 298-306
128. L. Falaschetti, L. Manoni, D. DiLeo, D. Pau, V. Tomaselli, C. Turchetti A CNN-based image detector for plant leaf diseases classification, HardwareX, 12 (2022), Article e00363
129. S. Sood, H. Singh A comparative study of grape crop disease classification using various transfer learning techniques, Multimed Tools Appl, 83 (2024), pp. 4359-4382
130. T. O'Halloran, G. Obaido, B. Otegbade, ID. Mienye A deep learning approach for Maize Lethal Necrosis and Maize Streak Virus disease detection, Machine Learning with Applications, 16 (2024), Article 100556
131. A.F. Rakib, R. Rahman, A.A. Razi, ASMT. Hasan A Lightweight Quantized CNN Model for Plant Disease Recognition, Arab J Sci Eng, 49 (2024), pp. 4097-4108
132. Y. Kaya, E. Gürsoy A novel multi-head CNN design to identify plant diseases using the fusion of RGB images Ecological Informatics, 75 (2023), Article 101998
133. M. Chilakalapudi, S. Jayachandran Optimized deep learning network for plant leaf disease segmentation and multi-classification using leaf images, Network: Computation in Neural Systems (2024), pp. 1-34
134. S. Verma, A. Chug, A.P. Singh, D. Singh Plant Disease Detection and Severity Assessment Using Image Processing and Deep Learning Techniques, SN COMPUT SCI, 5 (2023), p. 83

135. Iftikhar M, Kandhro IA, Kausar N, Kehar A, Uddin M, Dandoush A. Plant disease management: a fine-tuned enhanced CNN approach with mobile app integration for early detection and classification. Artif Intell Rev. 2024;57:167.
136. Sarfarazi S, Zefrehi HG, Toygar Ö. Potato leaf disease classification using fusion of multiple color spaces with weighted majority voting on deep learning architectures. Multimed Tools Appl. 2024. https://doi.org/10.1007/s11042-024-20173-3.
137. Gajjar R, Gajjar N, Thakor VJ, Patel NP, Ruparelia S. Real-time detection and identification of plant leaf diseases using convolutional neural networks on an embedded platform. Vis Comput. 2022;38:2923–38.
138. Ramamoorthy R, Saravana Kumar E, Naidu RChA, Shruthi K. Reliable and Accurate Plant Leaf Disease Detection with Treatment Suggestions Using Enhanced Deep Learning Techniques. SN COMPUT SCI. 2023;4:158.
139. Goluguri NVRR, Devi KS, Srinivasan P. Rice-net: an efficient artificial fish swarm optimization applied deep convolutional neural network model for identifying the Oryza sativa diseases. Neural Comput & Applic. 2021;33:5869–84.
140. Mururgan TK, Revanth P. Soil classification, crop prediction, and disease detection using ML and DL–"agro insights." J Plant Dis Prot. 2024;131:2161–79.
141. Deshpande R, Patidar H. Tomato plant leaf disease detection using generative adversarial network and deep convolutional neural network. The Imaging Science Journal. 2022;70:1–9.
142. Thai H-T, Le K-H, Nguyen NL-T. Towards sustainable agriculture: A lightweight hybrid model and cloud-based collection of datasets for efficient

leaf disease detection. Future Generation Computer Systems. 2023;148:488–500.

143. Ledbin Vini S, Rathika P. TrioConvTomatoNet: A robust CNN architecture for fast and accurate tomato leaf disease classification for real time application. Scientia Horticulturae. 2024;330:113079.
144. Sujatha K, Reddy TK, Bhavani NPG, Ponmagal RS, Srividhya V, Janaki N. UGVs for Agri Spray with AI assisted Paddy Crop disease Identification. Procedia Computer Science. 2023;230:70–81.
145. Shafik W, Tufail A, De Silva Liyanage C, Apong RAAHM. Using transfer learning-based plant disease classification and detection for sustainable agriculture. BMC Plant Biol. 2024;24:136.
146. Prashanthi Bh, Krishna AVP, Rao ChM. LEViT- Leaf Disease identification and classification using an enhanced Vision transformers(ViT) model. Multimed Tools Appl. 2024. https://doi.org/10.1007/s11042-024-19866-6.
147. Bhowmik AC, Ahad MdT, Emon YR, Ahmed F, Song B, Li Y. A customised vision transformer for accurate detection and classification of Java Plum leaf disease. Smart Agricultural Technology. 2024;8:100500.
148. Hemalatha S, Jayachandran JJB. A Multitask Learning-Based Vision Transformer for Plant Disease Localization and Classification. Int J Comput Intell Syst. 2024;17:188.
149. Kunduracioglu I, Pacal I. Advancements in deep learning for accurate classification of grape leaves and diagnosis of grape diseases. J Plant Dis Prot. 2024;131:1061–80.
150. Giakoumoglou N, Kalogeropoulou E, Klaridopoulos C, Pechlivani EM, Christakakis P, Markellou E, et al. Early detection of Botrytis cinerea

symptoms using deep learning multi-spectral image segmentation. Smart Agricultural Technology. 2024;8:100481.

151. Mamba Kabala D, Hafiane A, Bobelin L, Canals R. Image-based crop disease detection with federated learning. Sci Rep. 2023;13:19220.
152. Yu S, Xie L, Huang Q. Inception convolutional vision transformers for plant disease identification. Internet of Things. 2023;21:100650.
153. Nikith BV, Keerthan NKS, Praneeth MS, Amrita DrT. Leaf Disease Detection and Classification. Procedia Computer Science. 2023;218:291–300.
154. Wang Y, Rajkumar Dhamodharan US, Sarwar N, Almalki FA, Naith QH, R S, et al. A hybrid approach for rice crop disease detection in agricultural IoT system. Discov Sustain. 2024;5:99.
155. Chug A, Bhatia A, Singh AP, Singh D. A novel framework for image-based plant disease detection using hybrid deep learning approach. Soft Comput. 2023;27:13613–38.
156. S AN, R H, S R, Gaur S, R S. A smart agriculturing IoT system for banana plants disease detection through inbuilt compressed sensing devices. Multimed Tools Appl. 2023;82:45519–39.
157. Wang Y, Rajkumar Dhamodharan US, Sarwar N, Almalki FA, Naith QH, R S, et al. A hybrid approach for rice crop disease detection in agricultural IoT system. Discov Sustain. 2024;5:99.
158. Sood S, Singh H. A comparative study of grape crop disease classification using various transfer learning techniques. Multimed Tools Appl. 2024;83:4359–82.
159. O'Halloran T, Obaido G, Otegbade B, Mienye ID. A deep learning approach for Maize Lethal Necrosis and Maize Streak Virus disease detection. Machine Learning with Applications. 2024;16:100556.

160. Hemalatha S, Jayachandran JJB. A Multitask Learning-Based Vision Transformer for Plant Disease Localization and Classification. Int J Comput Intell Syst. 2024;17:188.
161. Kunduracioglu I, Pacal I. Advancements in deep learning for accurate classification of grape leaves and diagnosis of grape diseases. J Plant Dis Prot. 2024;131:1061–80.
162. Bhowmik AC, Ahad MdT, Emon YR, Ahmed F, Song B, Li Y. A customised vision transformer for accurate detection and classification of Java Plum leaf disease. Smart Agricultural Technology. 2024;8:100500.
163. Hemalatha S, Jayachandran JJB. A Multitask Learning-Based Vision Transformer for Plant Disease Localization and Classification. Int J Comput Intell Syst. 2024;17:188.
164. Sujatha K, Reddy TK, Bhavani NPG, Ponmagal RS, Srividhya V, Janaki N. UGVs for Agri Spray with AI assisted Paddy Crop disease Identification. Procedia Computer Science. 2023;230:70–81.
165. O'Halloran T, Obaido G, Otegbade B, Mienye ID. A deep learning approach for Maize Lethal Necrosis and Maize Streak Virus disease detection. Machine Learning with Applications. 2024;16:100556.
166. Falaschetti L, Manoni L, Di Leo D, Pau D, Tomaselli V, Turchetti C. A CNN-based image detector for plant leaf diseases classification. HardwareX. 2022;12:e00363.
167. Nagachandrika B, Prasath R, Praveen Joe IR. An automatic classification framework for identifying type of plant leaf diseases using multi-scale feature fusion-based adaptive deep network. Biomedical Signal Processing and Control. 2024;95:106316.

168. Sahu K, Minz S. Adaptive fusion of K-means region growing with optimized deep features for enhanced LSTM-based multi-disease classification of plant leaves. Geocarto International. 2023;38:2178520.
169. Deeba K, Balakrishnan A, Kumar M, Ramana K, Venkata Narasimhulu C, Dhiman G. A disease monitoring system using multi-class capsule network for agricultural enhancement in muskmelon. Multimed Tools Appl. 2024;83:82905–24.
170. Xu Y, Mao Y, Li H, Sun L, Wang S, Li X, et al. A deep learning model for rapid classification of tea coal disease. Plant Methods. 2023;19:98.
171. Mishra AM, Kaur P, Singh MP, Singh SP. A self-supervised overlapped multiple weed and crop leaf segmentation approach under complex light condition. Multimed Tools Appl. 2024;83:68993–9018.
172. Chug A, Bhatia A, Singh AP, Singh D. A novel framework for image-based plant disease detection using hybrid deep learning approach. Soft Comput. 2023;27:13613–38.
173. Verma S, Chug A, Singh AP, Singh D. PDS-MCNet: a hybrid framework using MobileNetV2 with SiLU6 activation function and capsule networks for disease severity estimation in plants. Neural Comput & Applic. 2023;35:18641–64.
174. Kaushik I, Prakash N, Jain A. Plant disease detection using a depth-wise separable-based adaptive deep neural network. Multimed Tools Appl. 2024;84:887–915.
175. Abdallah HB, Henry CJ, Ramanna S. Plant species recognition with optimized 3D polynomial neural networks and variably overlapping time–coherent sliding window. Multimed Tools Appl. 2024;83:80667–700.

176. Upreti K, Singh P, Jain D, Pandey AK, Gupta A, Singh HR, et al. Progressive loss-aware fine-tuning stepwise learning with GAN augmentation for rice plant disease detection. Multimed Tools Appl. 2024;83:84565–88.
177. Xu Y, Mao Y, Li H, Sun L, Wang S, Li X, et al. A deep learning model for rapid classification of tea coal disease. Plant Methods. 2023;19:98.
178. Giakoumoglou N, Kalogeropoulou E, Klaridopoulos C, Pechlivani EM, Christakakis P, Markellou E, et al. Early detection of Botrytis cinerea symptoms using deep learning multi-spectral image segmentation. Smart Agricultural Technology. 2024;8:100481.

Chapter (3): The Role of Artificial Intelligence in Mitigating Agricultural Water Scarcity

Fuad M. Ahmad

Department of Soil and water, College of Agriculture Engineering Sciences, Salahaddin University-Erbil, Iraq.

Email: fuad.ahmad@su.edu.krd

Abstract:

Due to a number of reasons, including climate change, basin-level water availability, and management system flexibility, water scarcity and pollution are endangering the sustainability of water usage for residential and commercial purposes. Water scarcity is a complicated and dynamic problem. The demand for water for residential, commercial, agricultural, and recreational purposes will increase as economies and populations continue to grow, putting pressure on the finite supply of freshwater. Economic collapse and waterborne disease outbreaks brought on by inadequate sanitation could result from inadequate water management. Persistent water scarcity has led to disputes over commercial water rights in many parts of the world. To solve this difficulty, a number of solutions have been investigated, including desalination, rainfall harvesting, water loss reduction, smart water systems including irrigation, dew collecting, runoff capture, wastewater recycling, and the use of artificial intelligence (AI). The current state of water scarcity alleviation solutions is examined in this part, with a focus on artificial intelligence's (AI) role in long-term water supply sustainability.

Key words: Artificial Intelligence, Mitigating, Water Scarcity

Introduction

Artificial intelligence (AI) is essential in reducing agricultural water scarcity through the use of precision irrigation techniques. Decision-making can be aided by real-time data from sensors and remote sensing devices, which are sophisticated artificial intelligence technologies. Therefore, a variety of tasks, including learning, planning, problem-solving, and decision-making, can be replicated by robots. The water sector faces many challenges because of changing climatic patterns and increasing resource restrictions, including budgetary and manpower difficulties. The increase in extreme weather events has made water increasingly scarce, unpredictable, and dirty, which has put more pressure on water infrastructure, treatment facilities, and people in charge of managing it. The future of water and wastewater systems could be completely changed by AI technology, ensuring the industry's long-term viability.

Water quality monitoring

Artificial intelligence has the ability to constantly track the quality of water hence providing real-time data analysis. With the help of data that the sensors installed on the different bodies of water will capture, AI algorithms can detect the changes in the quality of the water and identify any possible pollutants or hazards to human health, such as pollution plumes and the spread of waterborne pathogens. Furthermore, AI technology will be able to identify the instances of eutrophication, harmful algal blooms, and illegitimate disposition of dangerous chemicals.

Leakage detection and prevention. Artificial intelligence has the potential to utilize the real-time data obtained with network sensors to measure, track, and control flow pressure and velocity to improve the energy consumption efficiency and decrease the operation expenses. Possibly with the help of advanced systems, hazardous sewage overflow may be prevented as much as technologies permit through the optimization of storage use in wastewater pump stations, pipes, and manholes. Also, predictive analytics can be implemented to identify abnormalities and predict, detect and rectify failures and blockages in the wastewater system. Artificial intelligence can also be used to speed up the notification to cleanup crews in cases where unexpected discharges happen. Infrastructure maintenance The artificial intelligence can be able to use modern data gathered by network sensors to evaluate, monitor, and enhance the flow pressure and velocity to contribute to the energy efficiency and reduce the operational costs. More advanced systems would help to prevent high risk sewage overflow during extreme weather conditions through changing the storage use in wastewater pump stations, pipelines, and manholes. Additionally, predictive analytics have the potential to reveal anomalies and predict, detect, and address the damages and blockages in the wastewater systems. Artificial intelligence can also help to speed up notifications to clean up teams in case of unexpected releases. Prediction and management of floods. Artificial intelligence has the potential of analysing the information gathered through various channels, including meteorological forecasts, river gauges and past flood data to predict and control flood instances successfully. AI can also be used to provide specific and timely notifications about floods by using advanced algorithms of machine learning, thus allowing authorities to introduce the necessary strategy to reduce the destruction inflicted by floods and safeguard their residents.

Water conservation

Artificial intelligence can advance tasks in using water by analysing the data that is collected by various sources such as weather trends, the moisture content of the soil, and the amount of water that crops need. With this data, AI algorithms can provide useful information on the timing and amount of water to use to irrigate crops and minimize unnecessary use of water to promote sustainable agriculture. 6.

Water resource management Water resources can be managed with the assistance of artificial intelligence, which can help better its overall efficiency. Through information gathered on the availability of the water, the consumption rates, and demographic expansion, the AI technologies can allow the governing structures to make informed decisions regarding the distribution of water and the designation of the associated infrastructure, especially in times of minimal supply.

In water resources, artificial intelligence (AI) and ChatGPT. The use of superior algorithms, including artificial neural networks (ANNs) and support vectors machines (SVMs), in water resources management has achieved significant improvements in the sphere of AI. At the outset, ANNs were commendable in the adjustment of river run-offs in regions that were mostly defined by a scarce historical data, by modeling complex non-linear interactions among climatic inputs and hydrological outputs. This advancement enhanced wider adaptation of AI to help solve hydrological problems especially those dealing with uncertainties of the changing patterns of rainfall and soil moisture where the conventional models have failed to cope with the limitation of data. The ability of AI to handle large volumes of data and predict extreme weather conditions (floods and droughts, etc.) has significantly enhanced disaster preparedness. In the Yangtze River Basin, AI-based flood risk prediction and reduction techniques have come in handy. Further, AI models have also been very useful with regard to groundwater management in predicting the groundwater levels and groundwater recharge in areas with limited water resources hence leading to more sustainable water management practices. Moreover, AI has been instrumental in the water quality management because real-time evaluation and monitoring of contaminants can be conducted by utilizing AI models which have been used to monitor the levels of turbidity in Ganges River. In the agricultural industry, the application of artificial intelligence (AI) to Smart

Microclimate Control Systems (SMCS) has revolutionized the use of resources in agriculture to maximize temperature, humidity, and soil moisture, leading to high crop productivity and better water use. These applications of AI in many different fields of the water resources management highlight the importance of the AI in dealing with the multifaceted problem of climate change, growing water demand, and environmental variability. As a result, AI has become an essential tool to improve the accuracy, efficiency, and sustainability of the water management processes. [1,2].

Artificial Intelligence (AI) allows the machine to imitate and, in some cases, surpass human mental abilities, providing them with the opportunity to quickly make rational choices and pursue certain goals, such as the production of multi-step predictions and the identification of patterns. AI is also capable of processing high amounts of data and handling complicated models without difficulty. The application of AI methodologies to various fields of engineering and science has continued to increase since the mid-1900s. The recent developments in AI with the help of interdisciplinary methodologies in the scientific approach have provoked a revolution in many fields, such as engineering, hydrology, technology, and medical imaging.

History of the use of AI in the management of water resources

The integration of artificial intelligence into the watershed management process has experienced a significant evolution over the past several decades driven by technological progress in the domain of AI and the growing complexity of the water-related issues. The use of AI in the field is not novel and began in the late twentieth century when the traditional deterministic or statistical-based hydrological models could no longer be effectively used to deal with the complex and dynamically changing nature of water systems. Such traditional models had to be reworked regularly and were not flexible enough to address changes in the environmental conditions at a fast pace caused by climate variability and human anthropogenic actions. Use of artificial intelligence in water resource management started in the early 1980s and went on until the early1990s, as machine learning started gaining prominence. Artificial Neural Networks (ANNs) were one of the earliest AI methods used in hydrology and had been demonstrated to be useful in forecasting complicated nonlinear relationships between climatic factors and hydrological responses including streamflow and groundwater levels. According to the initial research, ANNs were found to be more predictive than the conventional models in situations where basin data and records of past usage were few and thus allowing the network to discover complex nonlinear relationships between various variables. These achievements formed the base of the large-scale adoption of AI technologies when it comes to managing the domain of water resources.

In the 2000s, more sophisticated methods were added to the field with the introduction of the Support Vector Machines (SVMs) and fuzzy logic systems which were developed through the improvement of AI technologies. The improvements in these new techniques relative to the old ones were in dealing with uncertainties associated with hydrological processes, including the changes in rainfall distribution, changes on soil moisture content and the changes in land cover which would change water demands. The AI models started to provide viable solutions to the problem encountered by the traditional approach because of the drawbacks in data representation in the specific regions. The recognition that AI has surpassed the conventional statistical techniques in simulating complex systems was the historic point of water resource management that has increased the precision of forecasts and decision-making. [10]. The issues in water resources management and AI opportunities. Nowadays, water resources management is confronted with a great number of issues including climate change, population growth, and demand increase. Among the critical ones is the uncertainty of the meteorological event such as floods and drought, which cannot be forecasted correctly using the traditional statistical models. The application of artificial intelligence especially machine learning has improved the capacity to make more accurate predictions using a large amount of climate data thereby enhancing disaster preparedness and mitigation. Besides, the loss of groundwater sources, particularly in regions where the extraction is higher than sustainable, has become a problem of concern. The implementation of sustainable practices is possible due to the fact that the AI models, especially those that utilize the machine learning strategies, can predict the groundwater levels and recharge rates better than the traditional models. Another major issue is the pollution of the water quality by agricultural and industrial contaminants. Over the past years, the application of smart models to assess the water quality in real-time, estimate the possible contaminants, and locate the sources of pollution has gained more and more popularity as a viable contribution to the water quality management. [11-13].

Water conservation in these areas is a problem since there are conflicts between the needs of agriculture, industry, and urban populations. The use of AI in the process of making decisions also increases the effectiveness of water supply and infrastructure management to provide residents with more efficient solutions. These risks are further enhanced by the fact that aging water pipes are deteriorating, which can be solved by utilizing AI-based tools. Through the use of AI, future failures of pipes can be predicted ahead of time and hence, cost reduction through detection of faults at early stages and structural failure. [14, 15].

Perspectives of AI in enhancing hydrological processes. Artificial intelligence is significant to the water resources management since it is able to handle the complexities and dynamism, non-linear behaviour, and dynamic patterns of the natural hydrological systems. Such properties offer a complex of difficulties that are hard to realize and predict using traditional hydrological models, especially taking into consideration the impact of climate change and human activities. Even more efficient and accurate estimations of key hydrological variables and processes are provided by advanced AI methods, in particular, deep learning and neural networks. [16]. The use of AI models has significantly improved in most fields, such as prediction of rainfall, streamflow, and estimation of flood-related risks, particularly in limited data. In the recent researches, complex machine-learning models like Random Forest (RF) and Long Short-Memory (LSTM) have been used to forecast the intensity of rainfall, streamflow dynamics, groundwater recharge rates, and flood risks with better performance compared to the conventional models. Furthermore, the usage of Internet of Things (IoT) devices and remote-sensing technology has enabled the successful application of AI throughout the system of management using real-time data. The result of this integration has been the increase in crop-water supply, effectiveness of irrigation and flood-reduction. With the help of AI, water managers will be able to make informed decisions concerning the distribution of the resources at their disposal, especially in cases when it comes to extreme conditions such as natural disasters. Therefore, AI is instrumental in the processing and synthesis of large volumes of diverse data, which improves the knowledge of water systems and enables better management practices to be implemented. [17, 18]. Another reason why artificial intelligence is important in managing water-resources is that artificial intelligence can be applied to enhance sustainability. A combination of climate-projection data, land-use changes, and socio-economic data into AI models will allow obtaining an in-depth perspective on future water supply and demand. These models are important in the formulation of long-term water-management plans so as to accommodate availability despite the increasing unpredictability. [19].

Recent developments in machine learning in hydrological modeling and water resources in management. In the 2010s, a new trend in the field of water resources management is the implementation of artificial intelligence with the help of deep learning approaches. Machine learning, which is a component of AI, uses multilayer neural networks to identify trends in large sets of data. The long short-term memory (LSTM) is one of such techniques, which belongs to the group of recurrent neural networks (RNNs) and was explicitly created to predict time series. LSTMs have proved to be more effective in hydrology; they are used to forecast the streamflow, rainfall, and floods with different time intervals. The high-quality deep-learning-based models have significantly enhanced the accuracy and consistency of hydrological forecasting, especially in areas that have few historical data where the traditional models have proven to be problematic. The massive adoption of these models in various hydrological research highlights the paradigm shift that AI can bring to the disciplines of hydrology. In flood forecasting as well as risk management, artificial intelligence has significantly impacted practices in water management besides hydrological modeling. Advanced machine-learning models have been utilised namely, random forests (RF) and gradient-boosting machines (GBMs) to analyse large volumes of data of various sources including remote sensing data, weather, and Internet-of-Things sensors to effectively forecast the risk of floods. Such sophisticated AI models have been deployed in flood-prone areas like the Yangtze River Basin in China and the Ganges River Basin in India, and they give essential information regarding disaster preparedness and mitigation mechanism. A lot has been done in the application of artificial intelligence in the management of groundwater. Groundwater is an important resource especially in arid and semi arid regions whereby surface water is limited. The different AI techniques that have been employed to predict the groundwater levels, assess recharge rates, and model the effects of water extraction on the well-being of aquifers include artificial neural networks (ANNs) and hybrid models. Such predictive models have been particularly useful in the areas where there is water scarcity and have provided the much needed information in coming up with environmentally sustainable groundwater management plans that do not only co-exist with natural regeneration. Over the last couple of decades, artificial intelligence procedures have quickly attained mainstream as a means of dealing with the complex, dynamic, non linear, and volatile character of hydrological processes. The rising natural catastrophes, caused by climatic changes and global warming has brought grave difficulties to the hydrology and proper administration of water resources. As a result, the use of artificial intelligence models to process and predict hydrological changes has notably increased due to increasing uncertainty in hydro-geometeorological situations. The strategies of artificial intelligence offer promising additional or alternative to conventional physical or statistical methods of hydrological modelling.[30]

Artificial intelligence technologies are currently being used to develop accurate and strong hydrological models in finer spatiotemporal resolutions by using data on different sources like microsensing, imaging, in situ, and remote sensing devices. The development is important in learning intricate hydro-meteorological processes. Therefore, the research of new models of artificial intelligence is urgent to enhance the distribution, control, and conservation of water resources, which will eventually contribute greatly to their sustainable management. [31-35]. To the extrapolation and generalization of AI models in water resources. To a very large extent, AI is also relevant in the management of water resources by use of the learning data to make correct predictions and decisions in a new scenario. This ability has been useful especially in hydrology where water systems are affected by many factors including climate change, land use and human activities. The ability of AI models to scale performance to new conditions of various situations is critical in understanding the impact of new weather conditions, e.g., the number of heavy floods or extended droughts driven by climate change. [36, 37] As an example, it is possible to use AI models using historical streamflow data to predict future weather conditions on a hydrologic basis. The capability of AI to forecast many different outcomes including extreme floods or droughts, which are gaining momentum in the world thanks to the climate change, proves its performance in generalization under different conditions. It should be mentioned that the extrapolation and generalization capabilities of the AI in prediction and forecasting can be highly valued in terms of improving decision-making in the process of using water resources. With the implementation of AI in decision-support systems, water managers will have a chance to consider the pros and cons of alternatives, as well as create the strategies that will be resilient in the context of conflicting interests in the conditions expected in the future. This would play a very important role in ensuring the sustainability of water resources in a fast changing world. [38, 39]. The use of artificial intelligence methods in water resources and hydrology has been examined by many studies. These researches have paid attention to forecasting the natural events and the correlation between water resources and food security, so-called the W-F diagonal. In the recent past, some of the technologies that have been integrated in the field of AI in water resources and management have included remote sensing, GIS and IoT. The integration has resulted in a more complete and new water management system. Indicatively, remote sensing (satellite imaging) used together with ground sensors has improved flood forecasting and water quality evaluation, thus enabling real time decision making in case of emergency. [40-42].

More recently, the artificial neural network methods, especially Long Short Term Memory (LSTM) networks and Convolutional Neural Networks (CNNs), have widely been used due to their effectiveness in representing temporal and spatial associations of hydrological data. Such models have been effectively implemented to activities such as flood inundation mapping, prediction of ground water level and smart irrigation systems management. The ability of these models to fully model the dynamics of a water system has made artificial intelligence a needed tool in the management of water-resources. An emerging trend in literature is the development of hydrological models where artificial-intelligence methods are applied. The hybrid models aim to use the power of both artificial intelligence and physics-based models towards enhancing the validity and reliability of the predictions in relation to hydrological conditions in the variety of applications. Such models have been especially useful in the areas with variable hydrology, where AI implementation makes the subsequent projections more pragmatic[43, 44]. Figure 1 shows a bibliographic network which further crystallises the basis of these studies.

Fig. 1

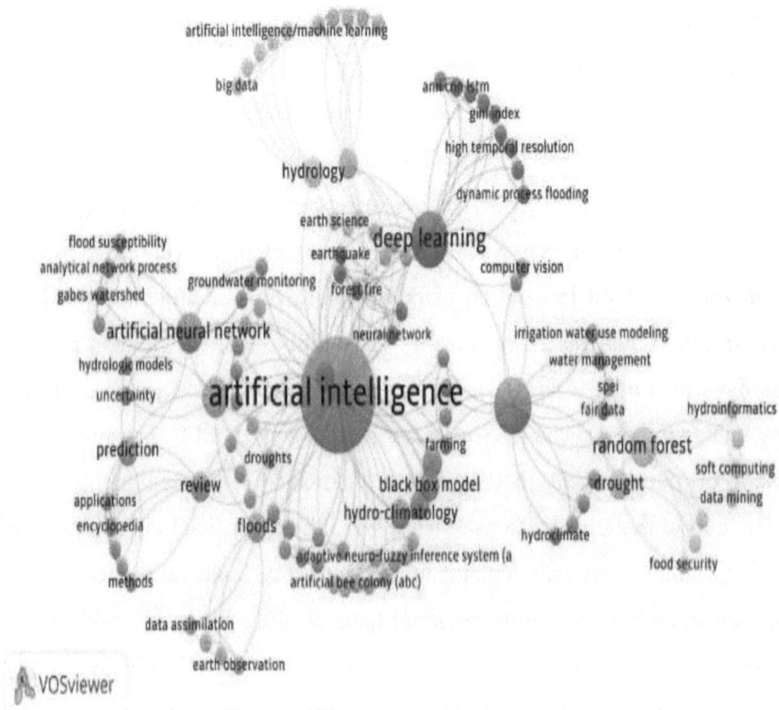

Bibliographic network of this study

Smart microclimate control systems enhanced by AI

Forecasting short-term microclimates presents considerable obstacles due to rapid changes and intricate relationships among climatic variables. To address this challenge, Chen et al. introduced an innovative microenvironment management framework with a specific focus on moisture control. This groundbreaking approach employs artificial training and system mechanics to regulate small regions within agricultural canopies, aiming to cool the atmosphere and enhance resource utilization. The application of this framework exhibits the feasibility of automation through neural networks in greenhouses, resulting in increased crop productivity and eco-friendly management of the intersection of agricultural, energy, and water resources. This integrated system utilizes real-time sensor data, historical climate data, and predicted weather patterns to monitor, predict, and manipulate various microclimatic elements such as temperature, humidity, moisture, light intensity, and CO2 levels. The core of the system consists of AI technologies including artificial intelligence (AI), machine learning (ML), reinforcement learning (RL), and fuzzy logic, with data analyzed by machine learning algorithms like neural networks to detect patterns and predict optimal conditions for plant cultivation. These models are instrumental in forecasting the appropriate humidity levels based on temperature and soil moisture, enabling precise environmental control. [45]. The system's control strategies are implemented through the utilization of the reinforcement learning (RL) algorithm. This approach enhances the decision-making processes involved in improving efficiency over time within the SMCS. For instance, the RL capability allows for the management of humidity and temperature settings, assimilation of feedback from the surrounding environment, and optimization of strategies for crop production. Moreover, fuzzy logic systems are employed to address the high uncertainty and complexity associated with environmental control and its interaction with various objectives, such as temperature and humidity. This capability enables the system to consistently make nuanced decisions that simplistic binary logic systems would struggle to execute effectively. One of the methods used to regulate the microclimate in the SMCS is through engineered ventilation, where the system utilizes temperature and humidity analysis features to automatically open or close greenhouse vents to achieve the desired microclimate.

Irrigation processes are also carefully controlled based on the current soil moisture content and weather patterns to ensure accurate water application without risking the detrimental effects of over or under-watering on crop yields. [46] Additionally, the AI models integrated into the SMCS have the capability to predict short-term changes in climate and proactively implement measures to mitigate potential issues. For example, when signs of high temperatures are detected within a building, the system will adjust ventilation and shading levels accordingly. Conversely, if low temperatures are detected, heating elements will be activated. This predictive capability also extends to resource management, such as water and energy usage, in order to optimize crop yields while promoting sustainability. To illustrate, the SMCS can be utilized in a tomato greenhouse to continuously monitor conditions and adjust variables based on forecasts. In the event of an anticipated temperature increase, the system will implement shading and improve air circulation to prevent overheating, while also regulating irrigation to prevent water wastage. By incorporating AI into microclimate management within the SMCS, a responsive system is created that adapts to real-time changes for enhanced agricultural outcomes and sustainable practices [47]. The utilization of the Smart Microclimate Control System (SMCS) in greenhouses exemplifies the application of AI in climate and resource management within the agricultural sector. Through the implementation of machine learning, reinforcement learning, and fuzzy logic, the SMCS effectively monitors variables such as temperature, humidity, and soil moisture to optimize crop yield. This system not only enhances the dependability of controlled environments but also offers viable AI solutions for open-field farming, including efficient irrigation, weather damage prediction, and resource distribution management. As a result, the increased integration of AI, as demonstrated by the SMCS model, influences various agricultural practices aimed at addressing climate variations and bolstering the sustainable nature of farming methods to ensure the resilience of food production systems. [48].

AI-powered environment scripting enabling intelligent metropolitan farming

Weather significantly influences crop growth in outdoor agricultural environments, highlighting the necessity of climate information for effective planning of cultivation and harvesting schedules, especially in urban farming. An independent mapping tool has been developed to analyze the climate trends in Taipei City, utilizing data from major meteorological factors across five climate observatories in Northwest Taiwan between 2014 and 2018. The outcomes of this analysis provide valuable insights for predicting future temperatures within specific time frames, enabling informed decisions on scheduling cultivation activities, making necessary adjustments, and optimizing the use of water and electricity to support environmentally sustainable farming practices in urban settings. Diverse weather conditions drastically affect crop growth and management practices, particularly in urban areas where climatic variations may differ significantly from surrounding rural regions due to factors like the urban heat island effect. These variations pose challenges to implementing standard agricultural methods, as traditional farming techniques do not always account for the relationship between structural elements and climate patterns. Addressing these challenges requires precise, short-term strategies that improve resource efficiency, productivity, and minimize environmental impact. Utilizing advanced, self-operating tools that can adapt to urban conditions can prove more effective in addressing these issues. An AI-powered mapping tool has been developed using machine learning algorithms to analyze and predict urban agricultural weather patterns with high spatial precision. This tool collects data on various weather aspects like temperature, humidity levels, precipitation, and wind patterns in different urban zones. This data, when processed through complex models, can generate accurate maps and forecasts tailored to microclimates within a city. Such forecasts are crucial for urban farmers, as they rely on precise climate information to make decisions related to planting schedules, irrigation management, and harvesting timings. [49, 50] In addition to controlling the consumption of water and electricity, the integration of the AI tool complements the broader goals of promoting ecological sustainability in urban agriculture. In addressing resource management challenges, the tool provides accurate location-specific forecasts that empower urban farmers to enhance efficiency and promote environmental conservation. For example, it can assist in determining the use of organic fertilizers or natural pest control methods by predicting favorable conditions for their effectiveness. This not only boosts crop yields, which traditionally rely on chemical inputs harmful to urban environments, but also contributes to overall environmental health. [51] Furthermore, because of the tool's predictive capabilities, it enables the coordination of farming operations across different areas of the city, fostering a cohesive approach to urban agriculture on a large scale. By aligning planting and harvesting schedules with projected weather patterns, the negative effects of unexpected climate variations can be minimized, resulting in a more consistent food supply within urban settings. This high level of synchronization is particularly beneficial in Community Supported Agriculture (CSA), wherein many small urban farms aim to meet the needs of local food consumers within their communities [46].

AI techniques in streamflow and rainfall-runoff modeling

Currently, Artificial Neural Networks (ANNs) are widely utilized for streamflow prediction because of their ability to effectively model nonlinear relationships between input variables such as rainfall, temperature, and land use, and output variables such as streamflow. ANNs must undergo a training process using historical data in order to make accurate predictions of future streamflow, thus facilitating applications in flood forecasting and water resources management. For instance, the use of ANNs in predicting the monthly streamflow of the River Nile has demonstrated superior accuracy compared to traditional methods, offering greater adaptability to seasonal variations and anomalies. [10, 22, 52].

Rainfall simulations

The escalation of climate change trends exacerbates the impacts of abnormal changes in precipitation and/or temperature, resulting in increased incidents, numerous environmental catastrophes, and severe storms. Despite advancements in empirical estimates of actual rainfall through climate monitoring, accurately assessing potential damage from rainfall remains challenging. Support Vector Machines (SVMs), an additional artificial intelligence technique, have been utilized for rainwater-runoff modeling. SVMs are particularly effective for analyzing small datasets and can efficiently and precisely model the rainfall-runoff process in areas with limited data availability. Research has shown that SVM-based models outperform linear regression models in predicting runoff in non-linear situations. eXtreme Gradient Boosting (XGBoost) has also been employed, along with a comparison to the primary method, XGBoost. By utilizing the XGBoost model, researchers were able to accurately determine the necessary amount of rainfall in un-irrigated basins, cross-referencing it with historical rainstorm events and subsequent damages to provide accurate predictions of flood-causing rainfall and recommend preventive measures for vulnerable areas. The educational approach utilizes Random Forest (RF) algorithms and Artificial Neural Networks (ANN) to simulate various scenarios of combined rainfall projections for southwest Asia. This method involved Modeling Aggregation Experiments conducted in North America, resulting in four prediction datasets—GEM-NEMO, NASA-GEOSS2S, CanCM4i, and COLA-RSMAS-CCSM4—all trained using data from the ERA5 revision. The results revealed that the RF and ANN techniques were notably superior as models compared to others, with RF demonstrating even greater dominance throughout the year, including leading time periods [53].

Support vector machines (SVMs) and random forests (RFs)

Support Vector Machines (SVMs) and Random Forests (RFs) are widely utilized Maximum Likelihood Estimators (MLEs) in rainfall simulation due to their ability to effectively handle the non-linear nature of input parameters like temperature, humidity, atmospheric pressure, and the output variable - rainfall amounts. SVMs are particularly suitable for situations with limited data and are designed to determine the optimal hyperplane that separates different classes, such as high and low rainfalls. On the other hand, RFs utilize a collection of decision trees to improve test error, minimize redundancy, and capture complex relationships within the data. SVMs and RFs have demonstrated satisfactory performance in various applications and have proven to be valuable tools in analyzing rainfall-runoff processes for water resource management. For instance, RFs have shown increased efficiency in predicting rainfall probabilities, especially in regions of the troposphere with unpredictable weather patterns. Despite their effectiveness, these methods may encounter challenges in scenarios where data streams are highly dynamic and non-stationary, a common occurrence in climate change scenarios [53].

eXtreme gradient boosting (XGBoost)

XGBoost has demonstrated superior effectiveness compared to both SVM and RF in replicating rainfall patterns. XGBoost represents an enhanced iteration of decision trees utilizing multiple concurrent tree models, with each successive model aiming to reduce errors from prior models. This intricate process results in a highly accurate model that is resilient to overfitting, particularly when analyzing extensive, multi-dimensional, and non-linear datasets. [54, 55] XGBoost has a notable advantage over other methods like SVMs and RFs, as it has the ability to effectively handle large amounts of data and can be adapted to various loss functions. Research has shown that XGBoost outperforms RF and SVM in predicting rainfall occurrences, especially demonstrating increased accuracy in ungauged basins with more limited data sets. For example, a recent study in mountainous regions concluded that XGBoost provided the most accurate estimates of rainfall runoff due to the complex topography and microclimates in the area [56].

Contributions to rainfall simulations

The advancement of artificial intelligence (AI) techniques has significantly improved the dependability and practicality of rainfall simulation. Support Vector Machines (SVMs) and Random Forests (RFs) play a crucial role in enhancing accuracy when data is scarce or when relationships are complex. However, research has revealed that XGBoost excels in scenarios requiring high precision in intricate environments, such as estimating rainfall in unmonitored basins or regions with erratic climates. The incorporation of XGBoost has empowered the algorithm to effectively capture interactions among various climatic factors, leading to more precise rainfall predictions. This enhancement plays a vital role in various sectors including water resource management, disaster planning, and agricultural decision-making where accurate rainfall estimates are imperative. [55].

Deep learning in hydrology

In recent times, alternative techniques such as Long Short-Term Memory (LSTM) networks have been utilized in the field of hydrology. The LSTM network, belonging to the Recurrent Neural Network category, has proven to be particularly beneficial for time series forecasting due to its ability to recognize prolonged patterns within the data set. Furthermore, LSTM models incorporating weather data have been successfully employed for the purpose of daily and monthly streamflow prediction. It has been noted that LSTM models excel in comparison to Artificial Neural Networks (ANNs) and Support Vector Machines (SVMs), particularly in accurately predicting peak flows and extreme flow conditions. [57]

Predicting weather

Various scientific studies highlight the significant role of environmental factors, particularly in relation to precipitation levels in streams. Researchers introduced a deterministic modeling technique to analyze daily precipitation patterns and their impact on extreme weather events. By employing a Markov model, they were able to predict rainfall occurrences on a daily basis and study the immediate influence of temperature fluctuations using a sophisticated predictive tool. The study was conducted in Spain's Jucar Creek Region, where a multidimensional probabilistic framework successfully replicated the occurrence of heavy rainfalls and temperature variations. The intricate nature of weather patterns presents challenges in conducting such research, particularly in the context of a constantly changing climate. Recent advancements in AI-supported models have improved the accuracy of weather forecasting, enabling more precise predictions. This section elaborates on how Markov models and multisite univariate self-correcting predictors are utilized in weather prediction, outlining the methodologies used to evaluate the effectiveness of each of these tools [47]

Markov models applied for weather forecasting

The process of weather prediction involves the utilization of statistical Markov models to analyze transitions in order to assess the probability of a system being in one state over another. These models are developed based on the concept of the Markov chain, where the subsequent state of the weather is determined by the current state, regardless of how it was attained. In practical applications, Markov models have shown promise in forecasting the likelihood of specific events, such as rainfall. In a particular area of study related to weather forecasting in the Mediterranean region, a Markov model was implemented to forecast daily precipitation occurrences. This model successfully captured seasonal variations and provided accurate short-term predictions, particularly benefiting regions with distinct weather patterns. [58, 59].

Multisite univariate self-correcting predictor

The enhanced AI methodology known as the multisite univariate self-correcting predictor addresses the challenges associated with the unpredictable and non-stationary nature of diverse weather patterns on a global scale. By incorporating data from multiple meteorological stations, this model continuously updates and recalculates its predictions based on real-time observations. This real-time correction feature ensures the accuracy of the model even when faced with sudden changes in weather conditions. In a study conducted in the Southern Alps, the multisite predictor was utilized experimentally to forecast extreme rainfall and temperature variations at different elevations. One of the key advantages of this model is its ability to adjust and self-correct based on incoming data from various locations. In comparison to traditional models, the multisite predictor demonstrated superior performance, particularly in regions with complex terrain where accuracy is crucial. [60].

Mitigating climate change impacts on agriculture through AI-driven

It is widely acknowledged that tackling climate change ought to be considered the most important challenge today. Climate change refers to the noticeable change in average weather patterns such as rainfall and warmth, averaged over long periods of time. In recent decades, however, it has been shown that active interference in the climate of the Earth's environment resulting in harmful global climate change can be halted due to people's aggressiveness. The increasing levels of CO_2 are predicted to result in crop fertilization, as the levels of energy needed in the warmer environment will be lower. While it could be said that climate change enhanced changes in the twentieth century affected countries in more of a positive light until 1980, these trends were further maintained by developed countries. This perspective then began to shift, such that latter developed nations began to bear these costs. In the 21st century, however, this is predicted to change so that developing as well as wealthy nations will sustain altercations by climate change.[61]

In the context of climate change agriculture is faced with the multifaceted problem of altered rainfall distribution, mean temperature, and extreme weather events, among others. Such factors as floods and droughts have a great deal of negative consequence on soil erosion. In addition, climate change modifies both the equilibrium of plant diseases and pests, as well as the amount of carbon dioxide and length of the growing seasons of crops. The other challenge is the rising sea levels. The temperature variation and the intensity of heat waves or cold snaps, the extremes of which crops and their yields are very sensitive to, are also a problem. The effect is determined by the stage of development of the plant in relation to the timing of the exposure to these extremes. Making matters worse, they aggravate the already difficult situation within cropping areas where water availability is limited, and rains are erratic, hence their irrigation strategies are unpleasant. Due to the increased uncertainty in rainfall amounts, farmers find it challenging to make productive decisions.[61]

Artificial Intelligence as we know it is revolutionizing many sectors, for instance we could mention the science of the improvement of cultivated plants. Artificial intelligence (AI) is widely recognized for its exceptional ability to examine and identify patterns within large datasets. We explore the wide potential of AI technology in various aspects of plant breeding, including data collection, exploring new genetic materials preserved in seed storage, and plants producing try to identify the relationship between structure and phenotype to expedite the development of draught crops. This approach will allow for the creation of crop varieties that are well-suited to the anticipated climatic conditions of the future.[61-63]

Additionally, these AI tools can enhance crop traits by enhancing the precision of gene editing technology to better predict the effect of specific genetic mutations on the associated plant traits. In addition, due to the advancement of AI assisted breeding methods, it will enhance the efficiency of precision breeding focusing on achieving localized optimizations for farming. They can plan better combinations for planting multiple crops in close proximity to one another or for alternating crops to make agriculture more sustainable and boost scientific yield in real-world greenery context. Climate change effect the morpho-biochemical and molecular processes of economically important crop species. Therefore, the current review highlights the role of AI and ML for crop improvement under current climate change [62- 64].

Role of AI (artificial intelligence) for crop improvement

AI employs multiple methodologies, such as machine learning, image recognition, and language processing, to drive agricultural productivity and efficiency. Machine learning programs sort through huge sets of data to forecast weather, schedule irrigation for optimal results, and identify diseases in crops before they cause widespread damage. Image recognition systems can identify plant illness, infestation by pests, and nutritional deficiency through robotic monitoring systems.

AI systems also use machine learning techniques to learn from patterns in the data to enhance their performance over time. Artificial intelligence (AI) is being used to rapidly create new plant types by combining large-scale genetic and trait analysis with modern breeding method (Figure 2). [65-71]

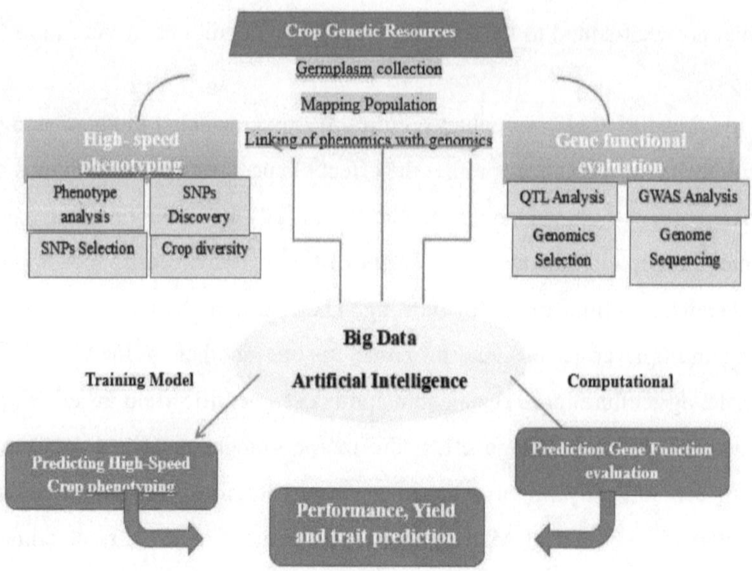

Figure2 Applications of AI and ML for Crop improvement.[69-71]

Artificial intelligence (AI) in crop development is also one of the key drivers and is used to exploit high-throughput phenotyping and gene functions. Procedural methods have been applied in order to efficiently and high-throughput acquire huge amounts of phenotypic and genetic information in large agricultural and breeding groups. This rich amount of data has formed the foundation of combining AI with different resources, such as phenotypic diversity of crops, SNP polymorphisms, QTL analysis, GWAS (whole-genome sequencing), genome selection, and genome sequencing. The use of AI systems is used to forecast phenotypes of crops through whole-genome sequencing, whereas AI-based calculations and model training enable the development of breeding approaches that are innovative. Therefore, the combination of AI and phenomics and genomics methods is likely to accelerate the discovery of genes that are relevant to agricultural phenotype and hence speeding up crop development efforts.[72-75]

AI-driven crop genomics and phenomics

Crop geneticists are striving to understand the genetic content involved in the plant phenotypes more, and examine the processes based on the molecular biology which generates such phenotypes, by using scientific data and bioinformatics. Artificial intelligence is a bridge into methodological strategies to solve a problem of unlike biological problems including metabolomics, proteomics, genomics, transcriptomics, and systems biology. The development of technologies made the process of the extraction of the genetic variation and observable traits by plants simpler and more accurate, which helped to extract valuable information out of the complex set of data. On the other hand, scientists have an opportunity to use artificial intelligence to study entire datasets of chemical processes that will allow explaining the molecular alterations that take place in non-living environmental strains.[76-78].

The transition of records between basic DNA and expressed bodily trends has been manipulated with the use of AI and the models that it has produced, allowing scientists to study the differences in the population of herbs. In particular, AI can be a helpful tool to breeders in the form of identification of genetic loci to stimulate the achievement of agricultural quality by the activation of the genetic algorithms, thus, aiding in the discipline of big-scale crop traits dimensions in both the laboratory and the controlled environment. Moreover, biological information analyses and genome sequencing can be easily subjected to AI to analyze a range of structural components of the molecules, including transcription factor binding sites. These components include extensive RNA which is not translated, small RNA-based regulators, genetic partnerships, protein-coding genes, selective sites of RNA molecules, and controlling regions of DNA[79- 82].

State of art crop breeding initiatives are gathering these enormous datasets and enabled them to harness the comprehensive diversity of genetic and trait data to discover novel genes related to particular traits. Artificial intelligence has also opened a new field of data analysis and calculations, thus, one can now analyze large volumes of data at a time with it. Furthermore, AI analysis to investigate the relationship between possible genes and regulatory factors is a new approach to discover neglected genes that could be important in the agricultural field[79- 83].
AI is also increasing the precision and speed at which crops grow, and as such, attracts a novel breeding approach and algorithms to build that are founded on data. It is based on this that we critically assess the role of genomic prediction models like mpsoyDNGP (Multi Parent soybean deep neural genomic prediction) in the crop breeding process. We talk about their prospects, their problems, and their current use. It is suggested that some strategies should be put in place to address the existing problems, including the improvement of parent selection, precise estimation of the joint influence of various traits and genes, formation of

interpretable deep learning models, and the factor of environment. As scientists expect to understand more about the abilities of AI in predicting genes, they are trying to expand the knowledge on AI in the agricultural sector especially in prediction of genomes. We also recommend that more research be carried out using artificial intelligence in a way that is sustainable and fair to the problems of the food system. Nevertheless, a significant amount of time spent in the conventional breeding process might be minimized with the help of the usage of AI methods. The AI models can be applied to forecast features of the crop that would have otherwise taken breeders many seasons to determine. Therefore, the breeding process will alter considerably, with AI no longer being restricted to accurate predictions based on a tremendous amount of genetic data[91, 92].

Evolution of genome prediction

The researchers expect to gain more insights on the potential of AI to be used in genetic prediction and hope to broaden the knowledge on the applications of AI in the agricultural industry, in particular, in the field of genomic predictions. Our policy is to support the further development of the systems of artificial intelligence through sustainable and fair application to the challenges of the food system. Genetic selection has received a tremendous breakthrough using deep-learning technologies. As an example, DeepGS (Deep Learning Genomic Selection) provides more accurate predictions of the phenotypic value, thus improving other traditional methods like rrBLUP. DNNGP is differentiated through the incorporation of multi-omics data of plants with the use of a deep neural network, along with its stacked hierarchical architecture in adaptive learning of features. The model can be a useful tool in managing breeding information on different scales, and the model has high predictive accuracy as well as it is much faster than standard models. Another application of the use of deep-learning technologies in genomic prediction in crop breeding is the recently developed SoyDNGP a sophisticated web server where convolutional neural network-based models are used to predict agronomic traits in soybean. Generally, SoyDNGP is a significant innovation in the study of genome-wide variation data [91-94].

AI for predicting climate impact on crop yield

Agriculture is strongly connected with the environment. Traditional methods of farming give significant contributions to greenhouse gas emissions, deforestation, soil erosion and water pollution. The environmental impact of farming can be reduced and a sustainable land and resource management can be achieved through the adoption of sustainable practices like conservation agriculture, agro forestry and precision farming. Climate change is defined as having more unpredictability. Farmers need to adjust in order to become more resilient to climate-related issues. The need to mitigate the losses and keep producing can be done through strategies such as planting of drought-resistant crops, ensuring efficient use of irrigation and implementing climate-smart technologies. Smart Agriculture (CSA) will be transforming the field with the introduction of Artificial Intelligence (AI). This new paradigm makes AI one of the tools of solution to the problem of climate change. This paper discusses the interaction between AI and CSA study with special attention to the potential of AI to complement adaptation measures in intelligent agricultural technologies. Agriculture contributes significantly to the GHG (Greenhouse Gas) emissions, mostly the livestock release of methane and the soil and fertilizer release of nitrous oxide. CSA is oriented on discouraging such emissions by encouraging the practice that will reduce GHG emission, and increase carbon sequestration in the soil and plants. These are accuracy in nutrient regulation, agro forestry, conservation agriculture and the use of renewable energy sources[95- 99]. Artificial intelligence systems are capable of analysing images of crops to determine diseases, nutrient deficiencies, and pests. Through computer vision and machine learning, AI will identify the patterns observed in the visual with references to crop health conditions, allowing farmers to intervene early in order to use specific treatment, which will lead to fewer losses of crops and more efficient resource utilisation. Precision agriculture includes AI that ensures the allocation of resources optimizes the use of resources and also increases the yields of crops. Algorithms of AI process the data of sensors, drones, and satellites to give data on the state of the soil, weather, and crop development. The farmers are able to regulate the irrigation, fertilisation, and pesticide use to particular parts of the field using this information, which reduces the amount of waste and enhances productivity as much as possible [100]. The autonomy of equipment will do a precise planting, harvesting, and weeding thus saving time and minimizing manual labour. Robots used in livestock feed, monitor, and milk, which provides the animal with the optimal and regular attention. AI helps in monitoring the health and welfare of livestock. This data is analysed by AI algorithms and allows detecting diseases at an early stage, predicting them, and optimizing feeding and breeding. This enhances animal welfare, high productivity, and reduction in veterinary costs. There is also the use of AI to monitor the environment, water quality, soil health, and biodiversity. AI algorithms process sensor data to determine risks to the environment and resource optimization. This helps to foster sustainable agriculture, minimizes environmental damages, and save resources in nature [100].

Improving drought resistance with AI (artificial intelligence). The agricultural challenges are being targeted through the use of safe solutions in a better way using advanced technologies like robotics, connected devices, neural networks, and machine learning. The technology enables to analyze images and monitor weather remotely as a result of which farmers can enhance the quality and raise harvest yields [101, 102]. Early diagnosis and prediction. AI algorithms may be used to detect drought early in its development using various data streams, such as weather data, satellite data to visualize the vegetation health, and even measurements of soil moisture. AI has a potential and it allows both the government authorities and farmers to take a more proactive stance in reducing negative impacts of drought by issuing warnings and creating precise predictions. After that, AI can also predict the seriousness and length of a drought, which will reduce the response time. This type of forecasting is indispensable in the context of drought management because it lays down a time frame within which actions to avert crop losses, conserve water and offer help to the impacted people can be pursued within the context of such forecasting capacity magnification [103]. Optimized water use The artificial intelligence software can enhance the management of water in agriculture by properly evaluating the content of soil moisture and the water needs of crops. There is also an opportunity of using AI to help farmers water their crops and to effectively irrigate their crops so that the amount of water is minimized and so that the perfect conditions are achieved during the dry season, leveraging the data[104]. Similarly to virtual water managers of farmers, AI can analyze data of moisture sold and climate, and elasticity of crop development to learn the exact amount of water needed by every crop. This allows farmers to plan the irrigation times correctly and apply a sufficient amount of water to each crop either in excess or deficit, which in turn prevents ill-effects on harvest or quality, [105].

Conclusions

Artificial intelligence (AI) is one of the tools that have greatly improved water resources management because complex tools like Artificial Neural Networks (ANNs), Support Vector Machines (SVMs), and hybrid models can be used.

Firstly, ANNs played an important role in improving river flow forecasts, particularly in areas where historically, only a small amount of data is available, by the ability to model complex non-linear interactions between climatic inputs and hydrological outputs. This achievement enabled wider usage of AI in the industry. With the increase in complexity of hydrological problems, AI-based approaches like SVMs, fuzzy logic systems became critical in removing the uncertainties, like the variation in rainfall pattern and the soil moisture content, where traditional models could not perform effectively due to the limitations of data. The disaster preparedness has greatly enhanced through the ability of AI to manage large volumes of data and predict severe meteorological conditions, including floods and droughts. As per the example, AI-enhanced flood risk forecasting and control tools in the Yangtze River Basin have been invaluable. The AI models have also demonstrated effectiveness in areas that lack water resources in groundwater management, as such as in regions that have scarce water resources, the models have been adopted to predict groundwater levels and rates of recharge which has enhanced more sustainable water management practices. Moreover, AI has had an important influence on the management of water quality by making it possible to continue assessing and monitoring contaminants in real time, which is proven by the fact that AI models are able to monitor the turbidity levels in the Ganges River. In the agricultural sector, AI-based Smart Microclimate Control Systems (SMCS) have changed the management of the resources in that the system can optimize temperature, humidity, and soil moisture which lead to increased crop production and improved use of water resources. All these different uses of AI in water resources management explain its importance in solving the complex problems of climate change, growing water demand, and environmental variability. As a result, AI has become an important instrument to improve the accuracy, efficiency and sustainability of water management practices. Recommendations The adoption of Artificial Intelligence (AI) in the sphere of water resources management has become one of the most important aspects, as it can bring significant benefits, but also introduce certain challenges. AI, including advanced applications like ChatGPT, has already shown a lot of potential in improving many fields of water management, including hydrological modelling, water quality assessment, and flood prediction. The ability of AI to work with huge amounts of data and detect more complex patterns has allowed making more accurate predictions and making the process of decision making more efficient. In particular, artificial neural networks (ANNs) and support vector machines (SVMs) have played a leading role in predicting streamflow and controlling groundwater in regions that have low data. Moreover, AI-based Smart Microclimate Control Systems (SMCS) have enhanced agricultural operations by controlling the resources like water, soil moisture, and temperature effectively and increased crop production and better conservation of resources. However, the concept of AI introduction into the water resources management is not flawless. An important issue is related to how well AI models can be scaled and used in various settings, and specifically in areas where data is scarce. Traditional hydrological models tend to fail in dealing with the issue of complexity and uncertainty of environmental data, which AI aims to address. However, AI models can vary significantly in their efficiency depending on the quality and quantity of data, and such variables might vary significantly between different regions. Moreover, interdisciplinary research should be continued in order to make the AI models more accurate and applicable to the real world. The ChatGPT as an artificial intelligence model can contribute to this field through assisting in spreading information, helping to analyze the data, and providing information about the issue of managing water in a complex way. It may help the stakeholders to understand the potentials and restrictions of AI and give directions concerning the application of AI-based solutions. However, there are also some issues with the use of ChatGPT, such as the accuracy and reliability of the information delivered, especially when it is used in highly important decision-making. Therefore, even though AI and ChatGPT have significant potential in improving the field of water resources management, it is vital to discuss the challenges related to their use. This involves the increase of AI models in scalability, the quality of data, and the growth of cooperation between AI specialists and water management professionals. With such measures, AI will still be able to help the practice of more sustainable and efficient water management practices, therefore, helping to curb the growing problems of climate change and increasing water demand.

References

1. Sun Z, Sandoval L, Crystal-Ornelas R, Mousavi SM, Wang J, Lin C, Cristea N, Tong D, Carande WH, Ma X, Rao Y. A review of earth artificial intelligence. Comput Geosci. 2022;159:105034. https://doi.org/10.1016/j.cageo.2022.105034.
2. Sharma P, Singh S, Sharma SD. Artificial neural network approach for hydrologic river flow time series forecasting. Agric Res. 2021;11(3):465–76. https://doi.org/10.1007/s40003-021-00585-5.
3. Kao IF, Liou JY, Lee MH, Chang FJ. Fusing stacked autoencoder and long short-term memory for regional multistep-ahead flood inundation forecasts. J Hydrol. 2021;598:126371. https://doi.org/10.1016/j.jhydrol.2021.126371.
4. Blake RW, Mathew R, George A, Papakostas N. Impact of artificial intelligence on engineering: past, present and future. Procedia CIRP. 2021;104:1728–33. https://doi.org/10.1016/j.procir.2021.11.291.

5. Shi F, Wang J, Shi J, Wu Z, Wang Q, Tang Z, He K, Shi Y, Shen D. Review of artificial intelligence techniques in imaging data acquisition, segmentation, and diagnosis for COVID-19. IEEE Rev Biomed Eng. 2021;14:4–15. https://doi.org/10.1109/rbme.2020.2987975.
6. Xie H, Chu HC, Hwang GJ. Trends and development in technology-enhanced adaptive/personalized learning: a systematic review of journal publications from 2007 to 2017. Comput Educ. 2019;140:103599. https://doi.org/10.1016/j.compedu.2019.103599.
7. Zhou Y, Guo S, Hong X, Chang FJ. Systematic impact assessment on inter-basin water transfer projects of the Hanjiang river basin in China. J Hydrol. 2017;553:584–95. https://doi.org/10.1016/j.jhydrol.2017.08.039.
8. Krysanova V, et al. How the performance of hydrological models relates to credibility of projections under climate change. Hydrol Sci J. 2018;63(5):696–720. https://doi.org/10.1080/02626667.2018.1446214.
9. Imrie CE, Durucan S, Korre A. River flow prediction using artificial neural networks: generalisation beyond the calibration range. J Hydrol. 2000;233(1):138–53. https://doi.org/10.1016/S0022-1694(00)00228-6.
10. Rajaee T, Ebrahimi H, Nourani V. A review of the artificial intelligence methods in groundwater level modeling. J Hydrol. 2019;572:336–51. https://doi.org/10.1016/j.jhydrol.2018.12.037.
11. Akhyar A, et al. Deep artificial intelligence applications for natural disaster management systems: a methodological review. Ecol Indic. 2024;163: 112067. https://doi.org/10.1016/j.ecolind.2024.112067.
12. Nordin NF, Mohd NS, Koting S, Ismail Z, Sherif M, El-Shafie A. Groundwater quality forecasting modelling using artificial intelligence: a

review. Groundw Sustain Dev. 2021;14:100643. https://doi.org/10.1016/j.gsd.2021.100643.
13. Lakshmikantha V, Hiriyannagowda A, Manjunath A, Patted A, Basavaiah J, Anthony AA. IoT based smart water quality monitoring system. Glob Transitions Proc. 2021;2(2):181–6. https://doi.org/10.1016/j.gltp.2021.08.062.
14. Ahmad S, et al. Water resources and their management in Pakistan: a critical analysis on challenges and implications. Water-Energy Nexus. 2023;6:137–50. https://doi.org/10.1016/j.wen.2023.10.001.
15. Fang J, Cheng X, Gai H, Lin S, Lou H. Development of machine learning algorithms for predicting internal corrosion of crude oil and natural gas pipelines. Comput Chem Eng. 2023;177: 108358. https://doi.org/10.1016/j.compchemeng.2023.108358.
16. Ibrahim KSMH, Huang YF, Ahmed AN, Koo CH, El-Shafie A. A review of the hybrid artificial intelligence and optimization modelling of hydrological streamflow forecasting. Alexandria Eng J. 2022;61(1):279–303. https://doi.org/10.1016/j.aej.2021.04.100.
17. Saleh MA, Rasel HM, Ray B. A comprehensive review towards resilient rainfall forecasting models using artificial intelligence techniques. Green Technol Sustain. 2024;2(3): 100104. https://doi.org/10.1016/j.grets.2024.100104.
18. Talaviya T, Shah D, Patel N, Yagnik H, Shah M. Implementation of artificial intelligence in agriculture for optimisation of irrigation and application of pesticides and herbicides. Artif Intell Agric. 2020;4:58–73. https://doi.org/10.1016/j.aiia.2020.04.002.
19. Sun Y, Nan Z, Yang W, Li L. Projecting China's future water footprints and water scarcity under socioeconomic and climate change pathways using an

integrated simulation approach. Clim Serv. 2023;30: 100385. https://doi.org/10.1016/j.cliser.2023.100385.

20. Khand K, Senay GB. Evaluation of streamflow predictions from LSTM models in water- and energy-limited regions in the United States. Mach Learn with Appl. 2024;16: 100551. https://doi.org/10.1016/j.mlwa.2024.100551.
21. Chu KS, Oh CH, Choi JR, Kim BS. Estimation of threshold rainfall in ungauged areas using machine learning. Water. 2022. https://doi.org/10.3390/w14060859.
22. Nourani V, Baghanam AH, Adamowski J, Kisi O. Applications of hybrid wavelet–artificial intelligence models in hydrology: a review. J Hydrol. 2014;514:358–77. https://doi.org/10.1016/j.jhydrol.2014.03.057.
23. Xiong M, et al. Identifying time-varying hydrological model parameters to improve simulation efficiency by the ensemble Kalman filter: a joint assimilation of streamflow and actual evapotranspiration. J Hydrol. 2019;568:758–68. https://doi.org/10.1016/j.jhydrol.2018.11.038.
24. Apaydin H, Sattari MT, Falsafian K, Prasad R. Artificial intelligence modelling integrated with singular spectral analysis and Seasonal-Trend decomposition using Loess approaches for streamflow predictions. J Hydrol. 2021;600:126506. https://doi.org/10.1016/j.jhydrol.2021.126506.
25. Nourani V, Baghanam AH, Adamowski J, Kisi O. Applications of hybrid wavelet–artificial intelligence models in hydrology: a review. J Hydrol. 2014;514:358–77. https://doi.org/10.1016/j.jhydrol.2014.03.057.
26. Chang FJ, Wang YC, Tsai WP. Modelling Intelligent Water Resources Allocation for Multi-users. Water Res Manag. 2016;30(4):1395–413. https://doi.org/10.1007/s11269-016-1229-6.

27. Chang LC, Chang FJ, Yang SN, Kao IF, Ku YY, Kuo CL, Amin IM. Building an intelligent hydroinformatics integration platform for regional flood inundation warning systems. Water. 2018;11(1):9. https://doi.org/10.3390/w11010009.
28. Jiang S, Zheng Y, Solomatine D. Improving AI system awareness of geoscience knowledge: symbiotic integration of physical approaches and deep learning. Geophys Res Lett. 2020. https://doi.org/10.1029/2020gl088229.
29. Tsai WP, Chang FJ, Chang LC, Herricks EE. AI techniques for optimizing multi-objective reservoir operation upon human and riverine ecosystem demands. J Hydrol. 2015;530:634–44. https://doi.org/10.1016/j.jhydrol.2015.10.024.
30. Zhou Y, Guo S, Chang FJ. Explore an evolutionary recurrent ANFIS for modelling multi-step-ahead flood forecasts. J Hydrol. 2019;570:343–55. https://doi.org/10.1016/j.jhydrol.2018.12.040.
31. Afan HA, El-shafie A, Mohtar WH, Yaseen ZM. Past, present and prospect of an Artificial Intelligence (AI) based model for sediment transport prediction. J Hydrol. 2016;541:902–13. https://doi.org/10.1016/j.jhydrol.2016.07.048.
32. Sihag P, Singh VP, Angelaki A, Kumar V, Sepahvand A, Golia E. Modelling of infiltration using artificial intelligence techniques in semi-arid Iran. Hydrol Sci J. 2019;64(13):1647–58. https://doi.org/10.1080/02626667.2019.1659965.
33. Singh B, Sihag P, Parsaie A, Angelaki A. Comparative analysis of artificial intelligence techniques for the prediction of infiltration process. Geol Ecol Landscapes. 2020;5(2):109–18. https://doi.org/10.1080/24749508.2020.1833641.
34. Xiong M, Liu P, Cheng L, Deng C, Gui Z, Zhang X, Liu Y. Identifying time-varying hydrological model parameters to improve simulation efficiency by

the ensemble Kalman filter: a joint assimilation of streamflow and actual evapotranspiration. J Hydrol. 2019;568:758–68. https://doi.org/10.1016/j.jhydrol.2018.11.038.

35. Yaseen ZM, El-Shafie A, Jaafar O, Afan HA, Sayl KN. Artificial intelligence based models for stream-flow forecasting: 2000–2015. J Hydrol. 2015;530:829–44. https://doi.org/10.1016/j.jhydrol.2015.10.038.
36. Krishnan SR, et al. Smart water resource management using artificial intelligence—a review. Sustainability. 2022. https://doi.org/10.3390/su142013384.
37. Adikari KE, Shrestha S, Ratnayake DT, Budhathoki A, Mohanasundaram S, Dailey MN. Evaluation of artificial intelligence models for flood and drought forecasting in arid and tropical regions. Environ Model Softw. 2021;144:1–29. https://doi.org/10.1016/j.envsoft.2021.105136.
38. Bennagi A, AlHousrya O, Cotfas DT, Cotfas PA. Comprehensive study of the artificial intelligence applied in renewable energy. Energy Strateg Rev. 2024;54: 101446. https://doi.org/10.1016/j.esr.2024.101446.
39. Abiodun OI, Jantan A, Omolara AE, Dada KV, Mohamed NA, Arshad H. State-of-the-art in artificial neural network applications: a survey. Heliyon. 2018;4(11): e00938. https://doi.org/10.1016/j.heliyon.2018.e00938.
40. Chang F-J, Chang L-C, Chen J-F. Artificial intelligence techniques in hydrology and water resources management. Water. 2023. https://doi.org/10.3390/w15101846.
41. Sun Z, et al. A review of earth artificial intelligence. Comput Geosci. 2022;159: 105034. https://doi.org/10.1016/j.cageo.2022.105034.

42. Schumann G, Giustarini L, Tarpanelli A, Jarihani B, Martinis S. Flood modeling and prediction using earth observation data. Surv Geophys. 2023;44(5):1553–78. https://doi.org/10.1007/s10712-022-09751-y.
43. Naeem K, Aloui S, Zghibi A, Mazzoni A, Triki C, Elomri A. A system dynamics approach to management of water resources in Qatar. Sustain Prod Consum. 2024;46:733–53. https://doi.org/10.1016/j.spc.2024.03.024.
44. Brookfield AE, Ajami H, Carroll RWH, Tague C, Sullivan PL, Condon LE. Recent advances in integrated hydrologic models: integration of new domains. J Hydrol. 2023;620: 129515. https://doi.org/10.1016/j.jhydrol.2023.129515.
45. Chen TH, Lee MH, Hsia IW, Hsu CH, Yao MH, Chang FJ. Develop a smart microclimate control system for greenhouses through system dynamics and machine learning techniques. Water. 2022;14(23):3941. https://doi.org/10.3390/w14233941.
46. Ghobadi F, Kang D. Multi-Step Ahead Probabilistic Forecasting of Daily Streamflow Using Bayesian Deep Learning: A Multiple Case Study. Water. 2022;14(22):3672. https://doi.org/10.3390/w14223672.
47. Hernández-Bedolla J, Solera A, Paredes-Arquiola J, Sanchez-Quispe ST, Domínguez-Sánchez C. A continuous multisite multivariate generator for daily temperature conditioned by precipitation occurrence. Water. 2022;14(21):3494. https://doi.org/10.3390/w14213494.
48. Chen T-H, Lee M-H, Hsia I-W, Hsu C-H, Yao M-H, Chang F-J. Develop a smart microclimate control system for greenhouses through system dynamics and machine learning techniques. Water. 2022. https://doi.org/10.3390/w14233941.

49. Huang A, Chang FJ. Using a self-organizing map to explore local weather features for smart urban agriculture in northern Taiwan. Water. 2021;13(23):3457. https://doi.org/10.3390/w13233457.
50. Li J, Li S, Lei J, Zhang X, Qi J, Tohti B, Duan Z. Analysis of spatial structure in the Kashgar metropolitan area, China. Land. 2022;11(6):823. https://doi.org/10.3390/land11060823.
51. Sishodia RP, Ray RL, Singh SK. Applications of remote sensing in precision agriculture: a review. Remote Sens. 2020;12(19):3136. https://doi.org/10.3390/rs12193136.
52. Nourani V, Paknezhad NJ, Tanaka H. Prediction interval estimation methods for artificial neural network (ANN)-based modeling of the hydro-climatic processes, a review. Sustainability. 2021. https://doi.org/10.3390/su13041633.
53. Pakdaman M, Babaeian I, Bouwer LM. Improved monthly and seasonal multi-model ensemble precipitation forecasts in Southwest Asia using machine learning algorithms. Water. 2022;14(17):2632. https://doi.org/10.3390/w14172632.
54. Vogeti RK, Mishra BR, Raju KS. Machine learning algorithms for streamflow forecasting of Lower Godavari Basin. H2Open J. 2022;5(4):670–85. https://doi.org/10.2166/h2oj.2022.240.
55. Mushtaq H, Akhtar T, Hashmi MZ, Masood A, Saeed F, "Hydrologic Interpretation of Machine Learning Models for 10-daily streamflow simulation in Climate sensitive Upper Indus Catchments," vol. 20, no. September, 2022, ///C:/Users/Rashid/AppData/Local/Mendeley Ltd./Mendeley Desktop/Downloaded/Mushtaq et al. - 2022 - Hydrologic Interpretation of Machine Learning Models for 10-daily streamflow simulation in Climate sensitive Upp.pdf

56. Ghobadi F, Kang D. Multi-step ahead probabilistic forecasting of daily streamflow using bayesian deep learning: a multiple case study. Water. 2022. https://doi.org/10.3390/w14223672.
57. Jiang S, Zheng Y, Solomatine D. Improving AI system awareness of geoscience knowledge: symbiotic integration of physical approaches and deep learning. Geophys Res Lett. 2020;47(13):e2020GL088229. https://doi.org/10.1029/2020GL088229.
58. Yutong X. Applications of Markov chain in forecast. J Phys Conf Ser. 2021. https://doi.org/10.1088/1742-6596/1848/1/012061.
59. Hernández-Bedolla J, Solera A, Paredes-Arquiola J, Sanchez-Quispe ST, Domínguez-Sánchez C. A continuous multisite multivariate generator for daily temperature conditioned by precipitation occurrence. Water. 2022. https://doi.org/10.3390/w14213494.
60. Vallam P, Qin X. Multi-site rainfall simulation at tropical regions: a comparison of three types of generators: rainfall simulation at tropical regions. Meteorol Appl. 2016;23:425–37. https://doi.org/10.1002/met.1567.
61. Khan MHU, Wang S, Wang J, et al. Applications of artificial intelligence in climate-resilient smart-crop breeding. *Int J Mol Sci.* 2022;23(19):11156.
62. MWAIS 2018 Proceedings. Midwest (MWAIS). Association for Information Systems.
63. Konar A. Artificial intelligence and soft computing: behavioral and cognitive modeling of the human brain. *CRC Press.*
64. Farooq MA, Gao S, Hassan MA, et al. Artificial intelligence in plant breeding. *Trends Genet.* 2024;40(10):891–908.

65. Jarrahi MH. Artificial intelligence and the future of work: Human-AI symbiosis in organizational decision making. *Business Horizons*. 2018;61(4):577–586.

66. Harfouche AL, Jacobson DA, Kainer D, et al. Accelerating climate-resilient plant breeding by applying next-generation artificial intelligence. *Trends in Biotechnology*. 2019;37(11):1217–1235.

67. Li Y, Zhou X, Li G. Bridging natural language and graphical user interfaces. In *Human-Comp. Int. Series*. 2021.pp. 463–493.

68. Dwivedi YK, Hughes L, Ismagilova E, et al. Artificial Intelligence (AI): Multidisciplinary perspectives on emerging challenges, opportunities, and agenda for research, practice and policy. *International Journal of Information Management*. 2021;57:101994.

69. Esposito S, Carputo D, Cardi T, et al. Applications and trends of machine learning in genomics and phenomics for next-generation breeding. *Plants (Basel)*. 2019;9(1):34.

70. Reinoso-Peláez EL, Gianola D, González-Recio O. Genome-enabled prediction methods based on machine learning. *Methods Mol Biol*. 2022;2467:189–218.

71. Crossa J, Pérez-Rodríguez P, Cuevas J, et al. Genomic selection in plant breeding: Methods, models, and perspectives. *Trends Plant Sci*. 2017;22(11):961–975.

72. Mochida K, Koda S, Inoue K, et al. Computer vision-based phenotyping for improvement of plant productivity: a machine learning perspective. 2019;8(1):giy153.

73. Jung J, Maeda M, Chang A, et al. The potential of remote sensing and artificial intelligence as tools to improve the resilience of agriculture production systems. *Curr Opin Biotechnol.* 2021;70:15–22.
74. Caudai C, Galizia A, Geraci F, et al. AI applications in functional genomics. *Comput Struct Biotechnol J.* 2021;19:5762–5790.
75. Wang Z, Liu Y, Niu X. Application of artificial intelligence for improving early detection and prediction of therapeutic outcomes for gastric cancer in the era of precision oncology. *Semin Cancer Biol.* 2023;93:83–96.
76. Williamson HF, Brettschneider J, Caccamo M, et al. Data management challenges for artificial intelligence in plant and agricultural research. *F1000Research.* 2023;10:324.
77. Xu C, Jackson SA. Machine learning and complex biological data. *Genome Biol.* 2019;20(1):76.
78. Godwin ID, Rutkoski J, Varshney RK, et al. Technological perspectives for plant breeding. *Theor Appl Genet.*2019;132(3):555–557.
79. Mahood EH, Kruse LH, Moghe GD. Machine learning: a powerful tool for gene function prediction in plants. *Appl Plant Sci.* 2020;8(7):e11376.
80. Long P, Zhang L, Huang B, et al. Integrating genome sequence and structural data for statistical learning to predict transcription factor binding sites. *Nucleic Acids Res.* 2020;48(22):12604–12617.
81. Gao P, Zhao H, Luo Z, et al. SoyDNGP: a web-accessible deep learning framework for genomic prediction in soybean breeding. *Brief Bioinform.* 2023;24(6):bbad349.
82. Ernst J, Kellis M. Chromatin-state discovery and genome annotation with ChromHMM. *Nat Protoc.* 2017;12(12):2478–2492.

83. Tong H, Nikoloski Z. Machine learning approaches for crop improvement: leveraging phenotypic and genotypic big data. *J Plant Physiol.* 2021;257:153354.

84. Shen Y, Zhou G, Liang C, et al. Omics-based interdisciplinarity is accelerating plant breeding. *Curr Opin Plant Biol.* 2022;66:102167.

85. Uzal LC, Grinblat GL, Namías R, et al. Seed-per-pod estimation for plant breeding using deep learning. *Computers and Electronics in Agriculture.* 2018;150:196–204.

86. Niedbała G, Piekutowska M, Weres J, et al. Application of artificial neural networks for yield modeling of winter rapeseed based on combined quantitative and qualitative data. *Agronomy.* 2019;9(12):781.

87. Niazian M, Sadat-Noori SA, Abdipour M. Artificial neural network and multiple regression analysis models to predict essential oil content of ajowan (*Carum copticum L.*). *Journal of Applied Research on Medicinal and Aromatic Plants.* 2018;9:124–131.

88. Altuntaş Y, Cömert Z, Kocamaz AF. Identification of haploid and diploid maize seeds using convolutional neural networks and a transfer learning approach. *Computers and Electronics in Agriculture.* 2019;163:104874.

89. Montesinos-López OA, Martín-Vallejo J, Crossa J, et al. A benchmarking between deep learning, support vector machine and Bayesian threshold best linear unbiased prediction for predicting ordinal traits in plant breeding. *G3(Bethesda).* 2019;9(2):601–618.

90. Khaki S, Khalilzadeh Z, Wang L. Predicting yield performance of parents in plant breeding: A neural collaborative filtering approach. *PLoS One.* 2020;15(5):e0233382.

91. Feng W, Gao P, Wang X. AI Breeder: Genomic predictions for crop breeding. *New Crops*. 2024;1:100010.
92. Xu Y, Zhang X, Li H, et al. Smart breeding driven by big data, artificial intelligence, and integrated genomic-enviromic prediction. *Molecular Plant.* 2022;15(11):1664–1695.
93. Ma W, Qiu Z, Song J, et al. A deep convolutional neural network approach for predicting phenotypes from genotypes. *Planta*. 2018;248(5):1307–1318.
94. Wang K, Abid MA, Rasheed A, et al. DNNGP, a deep neural network-based method for genomic prediction using multi-omics data in plants. *Mol Plant.* 2023;16(1):279–293.
95. Eli-Chukwu NC. Applications of artificial intelligence in agriculture: A review. *Engineering, Technology and Applied Science Research.* 2019;9(4):4377–4383.
96. Yessengeldin B, Khussainova Z, Kurmanova A, et al. Exploitation of natural resources in Kazakhstan: judicial practice for foreign investment. *J East Asia Int Law*. 2019;12(1):169–179.
97. Ma X, Gryshova I, Khaustova V, et al. Assessment of the impact of scientific and technical activities on the economic growth of world countries. *Sustainability*. 2022;14(21):14350.
98. Wang J, Wang J, Zhang Y. Artificial intelligence in agriculture: applications, challenges, and perspectives. *J Integr Agric.* 2019;18(12):2774–2785.
99. Ramazanov S, Petrova M. Development management and forecasting in a green innovative economy based on the integral dynamics model in the conditions of "Industry - 4.0". *Ideas*. 2020;1(1):9–30.

100. Gryshova I, Balian A, Antonik I, et al. Artificial intelligence in climate smart agriculture: toward a sustainable farming future. *Access to Science, Business, Innovation in Digital Economy*. 2024;5(1):125–140.

101. Talaviya T, Shah D, Patel N, et al. Implementation of artificial intelligence in agriculture for optimisation of irrigation and application of pesticides and herbicides. *Artificial Intelligence in Agriculture*. 2020;4:58–73.

102. Ayed RB, Hanana M. Artificial intelligence to improve the food and agriculture sector. *Journal of Food Quality*. 2021:1–7.

103. Kwon J, Kim K, Jo Y, et al. Artificial intelligence assessment for early detection and prediction of renal impairment using electrocardiography. *Int Urol Nephrol*. 2022;54(10):2733–2744.

104. Xiang X, Li Q, Khan S, et al. Urban water resource management for sustainable environment planning using artificial intelligence techniques. *Environmental Impact Assessment Review*. 2021;86:106515.

105. Tariq R, Cetina-Quiñones A, Cardoso-Fernández V, et al. Artificial intelligence assisted technoeconomic optimization scenarios of hybrid energy systems for water management of an isolated community. *Sustainable Energy Technologies and Assessments*. 2021;48:101561.

Chapter (4): Integrating Artificial Intelligence into Veterinary Clinical Diagnosis

Eva A Ajaj

Internal and Preventive Medicine, Clinical Pathology, College of Veterinary Medicine, University of Mosul, Iraq

E-mail: evaaisser2012@uomosul.edu.iq

Abstract:

Over the recent years, the use of artificial intelligence (AI) has radically changed various fields and healthcare is among them. Specifically, AI has played an important role in enhancing human and animal clinical examinations, diagnoses, and treatments. Essential and effective care has become a reality in the veterinary field with the integration of AI, which is beneficial to both animals and their owners. The essay analyzes the role of AI in the revolution of veterinary medicine, which highlights its impact on clinical examinations, diagnoses, and animal treatment. AI sensors and devices can track the vital signs and behaviors of the animals in real-time and as a result, support quick identification of possible health issues. Wearables that have AI-based algorithms are capable of monitoring temperature, heart rate, and other diagnostic variables. Finally, AI will lead to co-existence between practicing veterinarians, commercial developers of AI platforms, and veterinary radiology researchers, making AI effective and useful in veterinary radiology more efficient and efficient, and ensuring the best possible patient care at any given time.

Key words: Artificial intelligence, Veterinary care , Animal Health , Applications

Introduction

The idea of incorporating artificial intelligence into the clinical diagnosis in the veterinary domain involves using machine learning algorithms to implement tasks, such as studying medical imaging techniques, such as X-rays and computed tomography, tracking the vital physiological parameters, and processing large volumes of information to identify patterns that are compatible with earlier disease identification and tailoring of treatment plans. Artificial intelligence can be considered as a decision-support device as it helps the veterinarians to increase diagnostic accuracy, efficiency as well as the ability to process intricate informational data, which in the end leads to the improvement of animal patient care. Artificial Intelligence (AI) refers to the ability of computerised systems or robot-controlled computers to execute activities that are otherwise associated with intelligent beings. In essence, AI is a representation of machine intelligence, which follows cognitive functions including learning and problem solving, which are apparent in humans and animals. AI originally was a field within the academic community as early as 1956, and has been experimenting with various approaches, such as simulating the brain, modelling human problem solving, logic, knowledge bases, and imitating human behaviour [1, 2]. Despite having been recognized as an academic field since 1956, AI, in the course of many years, was a relatively minor field of scientific inquiry with very limited practical uses until recently. A broad field, which is represented by companion animal health, population and public health, zoonotic diseases, and animal production, veterinary medicine reflects the impact of AI at the intersection of many scientific fields, including philosophy, mathematics, neuroscience, control theory, cybernetics, computer engineering, and data science. It is hoped that an overlapping of these two broad and transforming disciplines will bring to bear a mutually enhancing effect. The tremendous progress and opportunities of the development, implementation, and clinical embedding of AI in human and animal healthcare imply positive perspectives. In human medicine, AI has been useful in precision medicine, tailoring treatment plans based on genomic differences and other factors, in the development and discovery of drugs, where the DeepTox, a deep-learning-based algorithm, assesses the toxicology of drug molecules, and in image diagnosis and image guidance surgery involving computer vision. However, all these bright prospects also imply certain challenges, particularly connected with the cognition, interpretation, and adoption of such potent and innovative technology, due to the rapid rate and dynamism of science and business development products. [3].

Artificial Intelligence and clinical practice in veterinarianism

AI has the ability to perform various roles in veterinary clinical practice, therefore changing the provision of veterinary care and improving animal outcomes. As shown in Figure 1, radiomics, which is a new frontier of precision medicine, involves the use of advanced mathematical computations used to quantitatively analyse medical imaging. It relays that medical pictures carry data about disease-specific process which surpasses the possibilities of human eyesight analysis. Radiomics is the process of extracting many features of images and it converts them into you know, numbers. When these data are subdued to sophisticated statistical analyses, they become part and parcel of the development of the precision medicine (PPM) which is personalized. Radiomics can be applied to a wide variety of medical imaging methods, such as the MRI method, CT, PET, and ultrasound. Machine learning (ML) is an essential sub-domain of AI since it plays a central role of developing algorithms that can be trained on both available data to make accurate predictions of new data [4]. The application of AI to animal health (AH) enables the handling of very complicated problems like quantitative and predictive epidemiology, precision-based therapeutic solutions to both animals and humans, and host-pathogen interactions. AI can assist in the diagnosis of the disease, offer highly precise diagnoses with the minimum of errors, clarify the complex biological interactions, suggest solutions, and improve the risk assessment and interventions, which are specific to the risk. [5].

Figure 1.

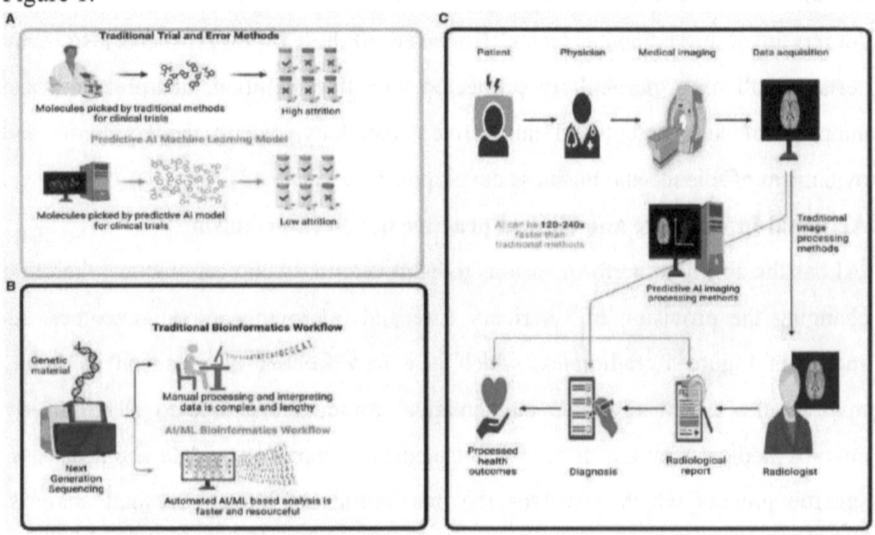

Diagrammatic illustration of the credible application of artificial intelligence (AI) and machine learning (ML) to the field of veterinary research.

The fields of animal health surveillance use AI and ML, such as predictive model building and finding viable drug targets to develop (A). ML may also be used in exploring data of genome sequencing (B). The use of ML in whole-genome sequencing can improve the source attribution, prediction of pathogenicity, prediction of antibiotic-resistance phenotypes and prediction of clinical outcome among veterinary patients. Radiomics, a very progressive facet of precision medicine, is a technique that uses sophisticated mathematical techniques to measure medical imaging quantitatively (C). Radiomics involves deriving a large number of features of images, which are translated into quantitative information. When such data are put to expert statistical applications, they will become part of advancing personalized precision medicine (PPM). Disease diagnosis Exploration of disease at an early stage is a necessary step to effective therapy, as it is possible to cause changes in patients and their prognosis. The use of AI in this scenario cannot be underestimated as it refers to the machine learning and sophisticated algorithms to process the vast amount of medical data, including the imaging and clinical data. The adoption of AI methods in the field of veterinary medicine is crucial in solving complex challenges in the fields of numerical and predictive epidemiology, precision medicine of people and animals, and host-pathogen interaction. The constant monitoring of at-risk patients based on AI-based systems will provide constant monitoring of a patient and make it easier to identify any changes in health indicators. This is more advantageous to chronic diseases, whereby it helps in intervening early and managing the patients proactively. Reagen did a study and found that chronic hypoadrenocorticism (CHA) in dogs was diagnosed by a machine-learning algorithm. Diagnosis of CHA is also a major problem because of the lack of awareness and index of suspicion. In the research conducted by Reagen, machine-learning methods were utilized to help diagnose CHA and used routinely obtained screening diagnostics, including complete blood count and serum chemistry panel [6]. The use of texture analysis (TA) has also been used to differentiate between non-infectious inflammatory meningoencephalitis and canine glial cell neoplasia. This activity can be extremely challenging to achieve mostly due to the overlapping image properties, even by the experienced diagnostic imaging experts, as evidenced in the study by Wanamaker et al., to which AI and radiomics are introduced as the tools to aid the decision-making process and integrate them into the routine clinical procedures and diagnostic, thus boosting sensitivity, precision, and reproducibility. Radiomics is a quantitative approach to medical imaging, which improves data interpretation with the help of advanced and even paradoxical mathematical analyses. [3, 7].

Zoonotic diseases surveillance

Artificial intelligence is one of the promising tools in monitoring and surveillance of zoonotic diseases. Specifically, the projected development of AI in the medical field and veterinary studies has great potential to overcome the challenges posed by newly developed zoonotic diseases. The combination of AI tools and traditional disease-control methods has presented new opportunities to comprehend, predict, and reduce the effect of the zoonotic pathogens by using the machine learning techniques and advanced algorithm models. There exists a great number of available AI predictive models that determine the risk variables and estimate a person on the likelihood of falling ill, which helps to identify them earlier and implement preventive measures. Both host-range susceptibility and viral host predication have been predicted using web servers and available AI models. As an example, Word2vec, a model in NLP, which vectors the viral nucleotide and protein sequences, predicts the host range of influenza virus. VIDHOP is a deep-learning approach that only predicts viral hosts based on viral nucleotide sequences only [812]. Web servers, like FluSPred (Flu Spread Prediction), a machine-learning derived tool, are used to predict human-associated strains of the Influenza A virus using protein and genomic sequence, implying whether a viral strain has the possibility of infecting human hosts. It was designed to focus on the high-risk viral strains to be studied further, facilitate the investigation into how an influenza virus emerges, and determine the risk that a novel influenza virus will have in case it gets the ability to be transmitted between people. [13].

Epidemiology Surveillance

Artificial intelligence can be implemented in line with the purpose of surveillance to improve the ability of research scientists and veterinarians to prioritize samples or cases with a high likelihood of positivity. This helps in distributing resources and laboratory facilities, hence maintaining focus on relevant specimens. The methodologies have been continually applied in animal disease surveillance such as food-borne diseases whereby metadata of the biological samples or cases are optimally used. Recent developments to detect and identify emerging animal diseases through machine-learning algorithms and web-based platforms like PADI-web and other bio-surveillance systems are an example of emerging technologies. Also, machine-learning methods have been used to infer genomic and epidemiological metadata and surveillance of food- and water-borne diseases simultaneously [14, 15]. Creation of Disease Models. The field of animal health surveillance is essential in machine learning (ML), especially during the creation of predictive models. One example is that the historical data of cases and risk factors can be used to identify farms with a higher propensity of infection by a particular pathogen. In Canada, trends in cases of porcine epidemic diarrhoea virus have been predicted by means of ML models. Host susceptibility mechanisms can also be explained by machine-learning methods. Becker et al. used ML to identify the potential disease reservoirs as well as the secret potential reservoir hosts of betacorona viruses, whereas Wardeh et al. used ML to determine the potential disease reservoirs. ML has also been applied to query genomic sequencing data; this data has complexity and hidden trends and poses an issue to traditional solutions, which is why ML has proven useful as a possible solution. The development of low-cost, high-throughput sequencing in veterinary and global health centers has enabled application of ML in whole-genome sequencing to enhance attribution of sources, evaluate pathogenicity, and predict phenotypes of antibiotic resistance as well as clinical outcomes. As a result, the improved knowledge of host-pathogen interactions and disease dynamics improves molecular surveillance systems of an extended variety of pathogens [16, 21].

AI in artificial Insemination

The large volumes of data required to study reproductive physiology, genetics, and the environment of an animal, artificial insemination with its massive capacity in data-processing is a key element in the analysis of large datasets. Through the absorption and processing of this information, AI algorithms can detect the best breeding periods, thus narrowing the list of important decision-making that can ensure the success of the artificial insemination procedures. Furthermore, artificial intelligence can be used to develop predictive models, which would consider multifactorial predictors of reproductive success. These models enable farmers and veterinarians to predict the timing of fertility, optimally choose breeding methods and perfect insemination procedures, hence improving reproductive success. Image-recognition technology based on AI is also instrumental in observing reproductive health, examining visual patterns like estrus behaviour, heat patterns, and general condition based on pictures and videos in order to better optimize on the timing and method of the insemination. Up to now, AI has transformed animal artificial insemination in that it has made it possible to identify prolific dams and animals with the highest prospects of success by using the historical data. AI has a significant effect of enhancing the effectiveness of reproduction processes; the accuracy that AI provides reduces the error of timing and method hence resulting in an increase in conception and general efficiency of reproductive processes in livestock breeding programs. Female infertile and sub-fertile males may be easily sorted out of breeding populations through use of semen analysis techniques including flow-cytometry-based analysis and computer assisted semen analysis (CASA). [22].

Assessment, treatment, and evaluation of the patients

Vital sign monitoring is a critical pre-diagnose study before diagnosis and treatment of disease in animals. Smart real-time monitoring systems with AI-powered analytical features make it possible to track the status of health indicators remotely, allowing veterinarians to track health statuses, detect the appearance of negative changes, and optimize treatment. The improvement facilitates in the choice of drugs, dosage calculation, and evaluation of the efficacy of different treatment methods. With the help of large-scale data analysis and modelling, AI systems are able to give veterinarians evidence-based recommendations and treatment courses that meet international standards. AI can increase accuracy in the decision-making process by suggesting individual treatment opportunities to a particular patient, based on scientific sources, similar cases, and the outcomes of treatment. Along with the inability to diagnose and make a prognosis without it, long-term monitoring of the disease is also paramount to review the results of therapeutic measures. In the dairy industry, machine learning has been examined, and especially in disease detection, e.g. mastitis and lameness, calving prediction, and milk output prediction [23-26]. AI analyses the individual health profile of any animal, considering the genetic tendency, age, breed, or medical history. With all this data combined, individual treatment programs that meet the needs and characteristics of each individual patient are created. The approach to pharmaceutical treatments optimizes the therapy of animals by evaluating their response to therapy and allowing veterinarians to tailor dosage, consider potential side effects, and even refine the treatment plan depending on the current data and the unique characteristics of each patient. AI has completely changed the game on the fight against antimicrobial resistance (AMR). It helps to identify the group of individuals at higher risk of contracting AMR, tracks the rise in popularity of resistant bacteria, detects the tendencies in the use of antibiotics, and quickly helps discover outbreaks of resistant infections. Introduction of AI in the assessment, treatment and evaluation of veterinary patients can transform the medical field of veterinary practice to be more accurate, efficient and personalized in the provision of care to the animal patients. The further development of the use of AI in veterinary practice will be refined and expanded through continuous research and cooperation between veterinary specialists and problems of AI specialists. [27].

Routine Surgery and Invasive Surgery

Computer enhanced visualization (CV) involves the process of acquiring and interpreting ultrasonic data of the imaging process using ultrasonic devices. Within techniques that use CV, fields that are essential in diagnosis and surgery are being transformed, especially image-based diagnostics and image-guided surgery (28). CV-enhanced imaging applications are becoming common in human and veterinary specialties, and include the X-ray, ultrasonography (US), computed tomography (CT), magnetic resonance imaging (MRI), positron emission tomography (PET) scans, retinal photography, dermo-scopey and histology, among others. One of the most noticeable issues that can be addressed with the help of AI and deep-learning methods is the subjectivity of image interpretation. Although many medical image analytics innovations have been reported in human fields over the past decades (28), their implementation in the veterinary clinical setting is still in its early stage, especially in low- and middle-income areas. A combination of radionic and artificial intelligence has the potential to improve several medical imaging applications, such as disease detection, characterization, and monitoring. A case in point is a study by Fraiwan and Abutarbush that proved the predictive power of such systems to identify the need to provide surgical intervention in a horse with colic as well as the ability to predict the result of surgical intervention. An example of an autonomous robot tissue is a smart tissue autonomous robot (STAR) that Johns Hopkins University has designed, which can handle similar procedures as those performed by human surgeons, such as bowel anastomosis in animals. The surgical systems with AI assistance will be able to elevate these rates considerably, as they will allow the creation of the operative algorithms that will recognize the best routes to take, assess the possible risks, and customize the surgical plans to fit the specific profile of the patient. These systems allow the surgeons to superimpose key information with real-time images, thus making work more precise, reducing risk of errors, and real-time prompts or advice that guide the decision-making process and direct the unexpected obstacles. [28, 29].

Limitations

As a rule, a few obstacles to the extensive implementation of artificial intelligence in the diagnosing and treating of veterinary medicine exist. Such obstacles are the lack of a robust reporting system, which would compel tens of millions of samples of images or text, many of which are not easily available; the organizational nature of the available samples, which is not always coded consistently or cohesively, thus hindering the creation of deep-learning models; the need to create deep-learning models based on labeled data; and the inability to label the data at the individual level, which is not yet done and necessitates organization with engineers and skilled workers. The information technology sector will have to make significant efforts to have the necessary efficacy, precision, and cost-efficacy. Besides, the implementation of electronic health records and their use require cooperation between laboratories, health centers, and veterinary facilities. Privacy laws are not applicable to the use of the patient data in the field of animal health care, but in human health care, strict data-privacy laws include the General Data Protection Regulation (GDPR) in the European Union and the Health Insurance Portability and Accountability Act (HIPAA) in the United States. These laws require storage, transmission, and access control of health data belonging to patients to be secure, which protects privacy and confidentiality. The algorithms of artificial intelligence in human health care are advantageous as they are based on systematically gathered and well-marked information that follows the FAIR Guiding Principles. Such systematic data collection is based on such standards as the Systematized Nomenclature of Medicine which has been adopted as the Systematized Nomenclature of Medicine-Clinical Terminology (SNOMED-CT), adopted by repositories of such data as the Veterinary Medical Database (VMDB) and the Small Animal Veterinary Surveillance Network (SAVSNET). However, the data may be difficult to retrieve and analyse since most of it is unstructured free text that cannot be readily queryable. [30-32].

The application of artificial intelligence in veterinary/biomedical research

Development of resistance in microbes is a significant health problem to the world especially in underdeveloped countries with poor healthcare facilities. Therefore, there is urgent need to come up with methods of fast production of new drugs, vaccines, and diagnostic devices. Artificial intelligence can have a significant impact on the antimicrobial resistance and the presence of antibiotic residues in food samples of meat and milk, as the current laboratory methods are time-consuming and prone to problems. The current dependence on the laboratory methods is tedious and fraught with a number of problems; therefore, artificial intelligence can be seen as the way forward to alleviate antimicrobial resistance to a great extent and to enhance the process of detecting antibiotic residues in food items like meat and milk. Artificial intelligence will be able to study large datasets to determine potential drug targets, as well as to develop new antibiotics, which possess increased activity and less resistance. Antimicrobial resistance research has been developed to use specific AI, such as Naive Bayes classifiers to predict antibiotic resistance and identify the main factors influencing resistance, decision trees to estimate the cost of resistance and to guide the use of antibiotics, random forest to predict the best antibiotic combinations, and support-vector machines to design new antibiotics. Antimicrobial susceptibility testing and whole genome sequencing innovative strategies have been developed using artificial intelligence. These new techniques are more accurate and quicker than the old techniques and have the potential of revolutionizing the diagnosis and treatment of antimicrobial resistance. An ideal development period of a new antibiotic would be ten to fifteen years but due to the emergence of artificial intelligence, the development period would be shortened to just an average of two years. This would be a major step in the direction of the research on antimicrobial resistance. The use of artificial intelligence in the detection of patients who are at high risk of contracting antimicrobial resistance infections, the spreading of resistant bacteria, antibiotic usage trends, and outbreak of resistant infections has transformed the fight against antimicrobial resistance. Consumption of antibiotic residues in food animals such as meat, eggs and milk leads to antibiotic resistance in human beings. Thus, the introduction of artificial intelligence in livestock production to identify and control the usage of antibiotics in animals will have an overwhelming effect on the fight against the threat of antibiotic resistance and will inform the researchers and scientists on the path of finding the trail of the transfer of resistance genes. Artificial intelligence can be used in case of food products, especially meat and dairy industry, to analyze food produced with antibiotic residues and trace back to their origin. Artificial intelligence is therefore a prospective tool with a potential of making significant contribution to antimicrobial resistance study in both livestock and human population and the future developments in AI technology will offer new solutions to deal with antimicrobial resistance. [34, 35].

Cancer research

Cancer is a dangerous condition that plagues human beings and animals. Although severe, therapeutic modalities, such as pharmacotherapy, radiotherapy, and surgery, have been highly investigated by researchers as potential treatments. The search of new ways by which cancer cells can be eliminated and the improvement of diagnostic techniques is hence essential.

Artificial intelligence is transforming veterinary oncology

Artificial intelligence-based tools play a critical role in enhancing the detection, control, and prevention of cancer. Several AI predictive models would be able to determine the risk factors and calculate the likelihood of a person developing malignancy. As a result, AI tools represent one of the promising opportunities in oncologic imaging that will significantly help radiologists to obtain accurate diagnoses. What is more, it is their important contribution to the discovery of oncologic drug targets. The instruments help in predicting the effectiveness and toxicity of candidate therapeutics and also improve treatment regimens. This efficiency is achieved by a deep analysis of large genomic and proteomic data sets, which demonstrates the wide and influential scope of AI in the development of the cancer research and treatment [39]. Development of such is a massive step forward, given that it has the potential to hasten the drug discovery cycle and shed light on novel oncologic drug targets that standard methodologies can miss. It has also made the way to individualized treatment regimes which are specific to an individual patient. Analyzing the characteristics of their neoplasms, AI systems can predict the manner in which patients would react to various treatment approaches, which is a massive breakthrough towards the more effective and personalized approaches to treating cancer. These developments do promise a lot in terms of improving patient outcomes and reducing the burden of oncology. It is worth noting that several models have been developed to make it easy to detect cancer early. In this specific research, researchers used an algorithm to detect malignant growths on computed tomography images. It is an algorithm that is based on AI and uses radiomics, showing the ability to retrieve important data as presented in a medical image that can be difficult to discern by the human eye. This is an example of how AI can complement the accuracy and effectiveness of cancer diagnosis, especially in the early phases where early intervention can have a significant impact on the course of treatment. Despite becoming a useful method of lung cancer screening, false-positive rates remain a problem despite the low-dose computed tomography (LDCT). The use of AI-based technologies is being used to improve the accuracy of the lung cancer diagnosis and minimize the false-positives. These artificial intelligence technologies use robust algorithms and image-processing to offer better detection rates of possible pulmonary malignancies, thus, providing stable and more consistent screening results. A study that was published recently created a deep-learning model associated with predicting the risk of future lung cancer based on a single LDCT scan, called Sybil, and confirmed it. [40-42].

Research on genomic and vaccine development

AI is quickly emerging as an essential methodology in genomic research as it provides a complex set of tools to assist the analysis of large, complex genomic data and the sequential extraction of meaningful insights. Algorithms that are artificial, and more so those that apply machine learning methods, are useful in identifying patterns, relationships that might not be detected by more traditional methods, thus providing the researchers with the necessary ability to make new discoveries and speed up advances in the field. Variant calling and annotation is one of the major uses of AI in genomic research. The algorithm is capable of recognizing and describing single nucleotide polymorphisms (SNP) and insertions and deletions (indel) and other functional and structural variations. The ability is especially important to whole genome sequencing (WGS) data, which yield information that would require the use of sophisticated analytical instruments to identify and analyze genetic variants. AI is also vital in the ability to predict effects of variants, assess possible functional implications of genetic alterations [43]. The AI system can approximate the probability that a particular variant could be used to contribute to disease or other characteristics of interest by examining pattern differences among populations and associating variants with phenotypic characteristics. This kind of information can be very useful in ranking variants to be studied further regarding their functional processes and creating an individual approach to medicine. The application of AI to improve phenotype-genotype mapping, which is the association of genetic variation with a specific trait or disease, is also applied. This system is able to reveal complicated associations among variants and phenotypes, including the ones which are subtle, or more than two genes. Such ability is necessary to understand the genetic basis of complicated diseases and to establish effective diagnosis and treatment methods. Vaccine design and development research has a great role of AI. The production of SARS-CoV-2 vaccine based vaccines in a short time, the cause of the COVID-19 pandemic, has highlighted the transformative nature of AI in the science of vaccines. AI is a potent tool that can speed up the process of identifying, designing, and evaluating vaccine candidates and thus it is an element set to transform the battle against infectious diseases. AI has the capacity to process large biological data, such as genomic and proteomic data, to determine which antigens have a high propensity to be immunogenic in the body [43, 44]. It is also able to predict the binding of affinities between antigens and immune molecules, which can help it to choose antigens that are effective at stimulating a protection effect. The AI-powered tools can design the vaccines, such as selection of adjuvants and other materials that enhance immunogenicity, and analyse intricate data to identify adjuvants, which interacts synergistically with antigens to enhance vaccine efficacy. Predicting efficacy and safety of potential vaccine candidates before clinical trials and analyzing historical data to reveal patterns correlating with efficacy and safety are some of the applications of AI. These predictions allow eliminating potential candidates in terms of further development, which will save time and other resources spent on testing the ineffective or risky vaccines. Research on genomic and vaccine development. AI is quickly emerging as an essential methodology in genomic research as it provides a complex set of tools to assist the analysis of large, complex genomic data and the sequential extraction of meaningful insights. Algorithms that are artificial, and more so those that apply machine learning methods, are useful in identifying patterns, relationships that might not be detected by more traditional methods, thus providing the researchers with the necessary ability to make new discoveries and speed up advances in the field. Variant calling and annotation are one of the major uses of AI in genomic research. The algorithm can recognize and describe single nucleotide polymorphisms (SNP) and insertions and deletions (indel) and other functional and structural variations. The ability is especially important to whole genome sequencing (WGS) data, which yields information that would require the use of sophisticated analytical instruments to identify and analyze genetic variants. AI is also vital in the ability to predict effects of variants, assess possible functional implications of genetic alterations [43]. The AI system can approximate the probability that a particular variant could be used to contribute to disease or other characteristics of interest by examining pattern differences among populations and associating variants with phenotypic characteristics. This kind of information can be very useful in ranking variants to be studied further regarding their functional processes and creating an individual approach to medicine. The application of AI to improve phenotype-genotype mapping, which is the association of genetic variation with a specific trait or disease, is also applied. This system is able to reveal complicated associations among variants and phenotypes, including the ones which are subtle, or more than two genes. Such ability is necessary to understand the genetic basis of complicated diseases and to establish effective diagnosis and treatment methods. Vaccine design and development research has a great role of AI. The production of SARS-CoV-2 vaccine-based vaccines in a short time, the cause of the COVID-19 pandemic, has highlighted the transformative nature of AI in the science of vaccines. AI is a potent tool that can speed up the process of identifying, designing, and evaluating vaccine candidates and thus it is an element set to transform the battle against infectious diseases. AI has the capacity to process large biological data, such as genomic and proteomic data, to determine which antigens have a high propensity to be immunogenic in the body [43, 44]. It is also able to predict the binding of affinities between antigens and immune molecules, which can help it to choose antigens that are effective at stimulating a protection effect. The AI-powered tools can design the vaccines, such as selection of adjuvants and other materials that enhance immunogenicity, and analyse intricate data to identify adjuvants, which interacts synergistically with antigens to enhance vaccine efficacy. Predicting efficacy and safety of potential vaccine candidates before clinical trials and analyzing historical data to reveal patterns correlating with efficacy and safety are some of the applications of AI. These predictions allow eliminating potential candidates in terms of further development, which will save time and other resources spent on testing the ineffective or risky vaccines.[45].

Phytomedicinal, ethnomedicinal and pharmaceutical studies. Artificial intelligence can be used in better drug discovery and the development of processes based on natural products. Innovative therapeutics development is necessarily complicated, time-consuming, and tedious. Some of the factors that restrain drug development include methods and techniques like trial-and-error experimentation and high-throughput screening. However, artificial intelligence is crucial in speeding up timelines by means of effective and efficient analysis of huge datasets. Artificial intelligence can discover and screen millions of potential drug candidates, thus saving a large portion of time and effort to design drugs; AI can also discover new drug targets that have therapeutic potential. AI could revolutionize the drug-discovery process by increasing the speed with which promising candidates are identified, optimizing clinical-trial design, and the possible opportunities of using AI to approach personalized medicine approaches. The AI models are able to forecast the effectiveness and possible undesirable outcomes of the candidates, allowing the researchers to focus on the most promising compounds to develop and, as a result, reduce the number of failed clinical trials. Besides, AI enables the creation of personalized medicine, which is the process of creating drug medications based on the specifics of the patient to achieve a maximum therapeutic effect and reduce side effects. The clinical-trial design can also be optimized through AI that will identify the most appropriate patient groups, treatment interventions, and outcomes, which will result in an efficient and effective development process.

Artificial intelligent diagnostics in veterinary medicine in a multimodal environment

This functionality has a special promise in veterinary diagnostics because animal health care is a complicated practice with many possible sources of data, such as medical imaging (e.g., radiographs, ultrasound), videotapes of animal behavior, lab results, and physical examination records. Multimodal AI enhances a more comprehensive view of the health of the patient by correlating these inputs and has been shown to increase the diagnostic accuracy as well as clinical workload and efficiency in decision-making. The sphere of veterinary diagnostics experiences some unique issues that are not common in the practice of human medicine. The wide range of species brings a lot of physiological and anatomical diversity that has implications on diagnostic thresholds and model generalizability. What is more, the fact that animal patients do not speak increases the significance of objective and data-driven interpretations based on various modalities. Lameness, respiratory illness are conditions which demand fine-grained analysis of biomechanical and acoustic patterns that can vary highly across species. However, multimodal AI has not yet been fully applied in veterinary medicine despite its potential. Even the existing diagnostic processes are based on unimodal process, with little intermodal integration. The current reviews tend to consider multimodal systems as peripheral or specific applications. [56].

Multimodal AI as a Veterinary Diagnostic Theoretical Underpinning

The multimodal artificial intelligence that has been applied in the context of veterinary diagnostics is based on the notion that combined heterogeneous types of data (imaging studies, text, auditory recordings and behavioral video) may provide more robust and comprehensive information than unimodal systems. This hypothetical basis has been termed as the exploitation of data complementarity allowing the seizure of both modality-specific features as well as cross-modal associations that increase diagnostic functionality and clinical insight. Three categories are used to generally describe multimodal fusion strategies: early fusion, late fusion and hybrid fusion. Early fusion: This is related to learning joint representations since at the beginning of the learning, the model combines data at the feature level. Modality -specific models are fused together in the late fusion which allows a modular design of the system. Hybrid fusion combines the two methods and allows more comprehensive and versatile integration between modalities. Contrary to human patients, animals offer a rich spectrum of baselines of vital signs, morphology of organs and behaviour. Additionally, the lack of linguistic skills in animals to express the symptoms makes the use of indirect ones (e.g. posture, gait, sounds of respiration or clinical notes) more essential, each of which do not exist in every species. This and other reasons make multimodal fusion strategies vital to design well and apply to the veterinary field. An example is that gait analysis among horses needs alternative temporal modeling than in small animals and the interpretation of respiratory sounds has to consider-species-specific acoustic patterns. Multimodal AI in veterinary diagnostics is currently in the context of integration and combination of imaging, behavioural analysis, clinical notes and laboratory data. It is within the bounds of recent developments in machine learning and deep learning methods, and their applications, difficulties and emerging tendencies. Figure 2 shows the methodology used.

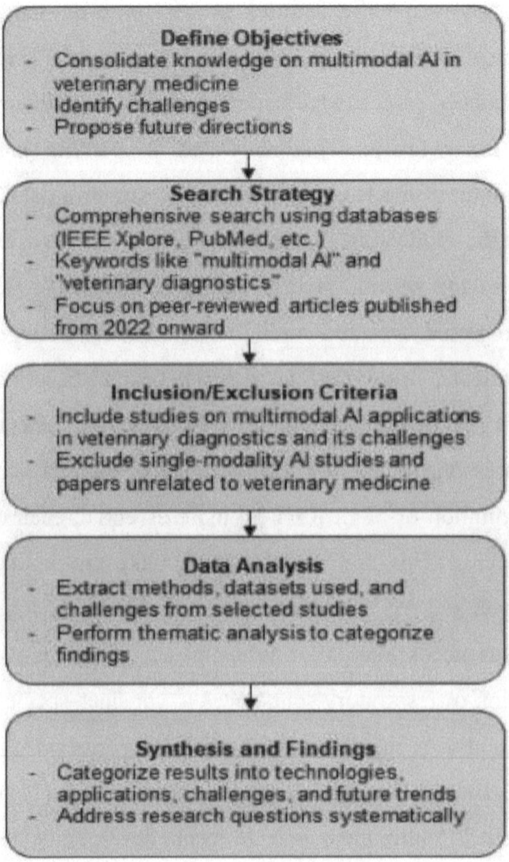

FIGURE 2.

Multimodal AI in veterinary diagnostics

Multimodal AI is used in Data Modalities. Multimodal AI is an AI that uses sophisticated computational methods to combine and process various types of data (Figure 3) to design a single system of decision-making. This process is similar to the holistic diagnostic method regularly used in veterinary medicine where data in many forms, including imaging, behavioral data, and clinical notes, is integrated to create a unified picture about the health of an animal patient. Recent efforts in multimodal AI in human medicine have been directed at making multimodal diagnostic frameworks, i.e., imaging, text, and audio modalities, to form a complete diagnostic decision-making process. These systems apply sophisticated approaches to multimodal fusion and analysis, which make it possible to have better contextual insight and predictive accuracy. An example of multimodal AI in human healthcare is a study that has been able to integrate imaging data with patient records to detect cancer and cardiovascular diseases [56- 58].

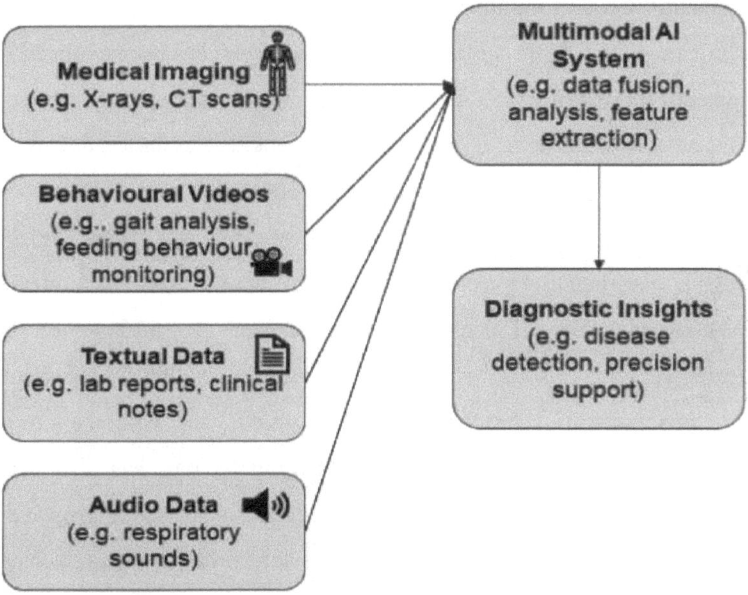

FIGURE 3. Multimodal AI integration in veterinary diagnostics

Medical Imaging Multimodal artificial intelligence systems are used to examine medical imaging with the implementation of Convolutional Neural Networks (CNNs) in order to identify patterns, abnormalities, and other diagnostic features. These models have been used in veterinary radiology to calculate fractures, tumors, or structural defects of X-ray and computed tomography. To exemplify, the application of radiographic systems based on AI to increase diagnostic efficiency and accuracy has been proven to be especially effective in the case of veterinary practice. These systems do not focus on the analysis of imaging alone but combine data from other modalities. As a case in point, the merging of patient history and imaging information through multimodal AI to improve diagnostic accuracy and give a more holistic picture of the situation with animal health has been explored in veterinary applications. Equally, in human medicine, multimodal methods are used to combine datasets, with the focus on automated segmentation of multimodal data, i.e., the combination of serial imaging with other clinical data to enhance the comprehension of complex diagnostic situations [59,-61]. Nevertheless, multimodal AI is also being developed in veterinary applications. A reference illustrating AI improvement of decision-making showed the analysis of the medical histories and comparisons with the imaging information, which transforms the diagnostic processes in veterinary hospitals. Besides, the combination of whole-slide imaging and clinical notes in veterinary pathology has also been mentioned, which enables more accurate diagnostic results in the animal. However, this synthesis is especially vital in the case of veterinary procedures, as species-specific physiological considerations and diagnostic peculiarities are present [62-64]. Behavioral Videos Behavioral videos provide dynamic information, making a record of movement or breathing, and thus giving more information about the health of the animal. An example is that gait-analysis systems can determine whether an animal is lame or has a neurological condition by examining the pattern of walking or running of an animal. Reference equine asthma on video recordings of the abdomen and nostrils was classified. Likewise, Reference performed an extensive survey of computer-vision techniques that have been used to identify behaviour patterns in livestock, pointing to its use in movement pattern tracking and the detection of abnormal situations in veterinary settings [65-67]. Motion estimation networks are often used for analogous purposes in the field of human medicine. Indicatively, Reference showed how multimodal AI could be applied to identify neurological diseases, including a Parkinson's disease, by analyzing video data. Although there are differences in application, the technology behind the usage is similar to that applied in veterinary care. The reference discussed how the video and accelerometer data can be used in the detection of lameness in animals and thus demonstrated the role of integrating wearable sensors with video-based motion analysis in animal diagnostics. However, specific needs of veterinary diagnostics, including the variety of species and their characteristic movements, remain the source of innovation in this field [6768]. Textual Data Textual records, such as laboratory reports, medical history, and clinical observations, provide the necessary background of the process of diagnosis. Natural Language Processing (NLP) methods have been used in veterinary medicine to produce meaningful information, trend detection, and correlation of results with other modalities. As an example, AI systems could be used in a veterinary to cross-check laboratory reports with imaging data to identify conditions such as diabetes or metabolic conditions in an animal. Reference consulted in a study aimed at investigating veterinary diagnostics, and discovered that combining textual patient narratives, lab reports, and imaging modalities could improve diagnostics in dermatology in companion animals [69]. NLP techniques have been applied in a broader way in human medicine. Ye et al. came up with a hybrid fusion method, which incorporates structured electronic health records (EHRs) and unstructured clinical notes to forecast patient outcomes. Although this type of methodology has mostly been studied in human situations, it can have veterinary application implications as well. On the same note, Reference suggested the use of deep-learning models on multimodal data, including textual EHRs and imaging scans, to enhance diagnostic accuracy on chronic diseases in humans. The potential of large language models to transform patterns implicit within large repositories of textual data, a method that has the potential to be applied to veterinary datasets, was also highlighted in this work. Reference showed that the multimodal AI systems used in human medicine are more interpretative when there is a correlation of imaging results and textual clinical observations. All these studies highlight the possibility of using the high-order NLP algorithms to reach more reliable and context-driven diagnostics in the veterinary and human fields. [70, 71].

Audio Data

In both veterinary and human medicine, audio data, e.g., respiratory sound recordings, are being increasingly used to increase diagnostic accuracy. Sound recognition algorithms are used in veterinary medicine to detect abnormalities like wheezing, crackles, and other respiratory illnesses indicators [72, 73]. In the field of human medicine, recent developments of audio-data integration have been directed towards predicting respiratory health. One of the references discussed the application of multimodal large language models (RespLLM) to combine audio and textual data and, thus, offer a more comprehensive view of the respiratory conditions in human patients. In addition, the research showed that acoustic data combined with clinical manifestation and imaging results enhances the accessibility and accuracy of resources in resource-constrained environments, mainly in human medicine. Although these methodologies are highly promising for human use, their adaptation to the veterinary environment provides special opportunities to increase diagnostic possibilities among the species [72, 77]. Multimodal AI systems provide more insight into respiratory conditions because they combine audio information with other modalities, including therapy images and patient records. Originally used in human medicine, the techniques are now also becoming equally groundbreaking when used in veterinary diagnostics and are being used to circumvent human-specific issues and broaden the application of real-time diagnostic technologies to the animal world.

Multimodal AI Multimodal AI obstacles to veterinary diagnosis

The implementation of multimodal artificial intelligence in veterinary diagnostics is faced with significant challenges, which include data collection, technical barriers, adoption barriers, and ethical issues (Figure 4). These problems need to be overcome in order to facilitate the successful implementation and wide adoption of such systems. The problems mentioned above are specific to the field of veterinary medicine, but they are related to similar barriers in human healthcare, which underscores the importance of interdisciplinary collaboration and innovation.

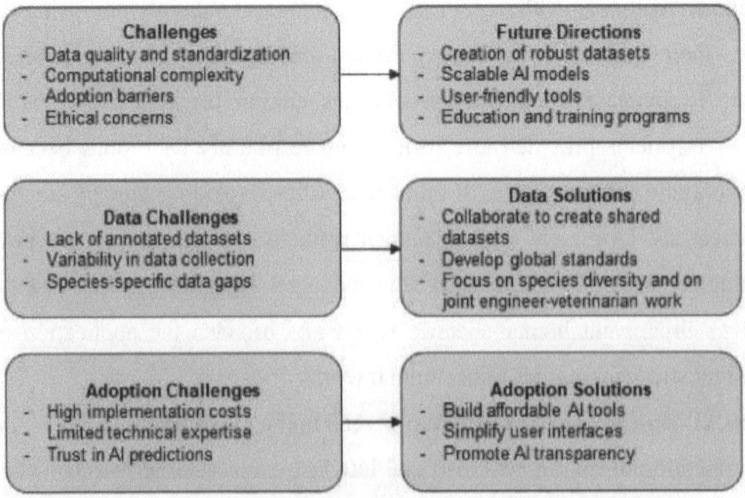

FIGURE 4. Challenges and solutions framework for multimodal AI

Data Collection and Quality One of the essential issues in veterinary medicine is the way to gather and ensure the quality of the data needed to train and validate multimodal AI systems. Quality annotated datasets do not exist in various modalities, including images, textual records, and behavioral data. Multimodal data is labor-intensive to annotate and may often need domain knowledge of veterinarians and technicians. This is further restricted by the fact that it is difficult to have standardization of data collected under different sources such as clinical notes, laboratory results, and imaging scans, which are likely to be in different formats, resolutions, and detail [79, 80]. Likewise, the difficulties in the standardization of datasets are also witnessed in human medicine when uniting multimodal datasets like electronic health records and imaging data. Nevertheless, in human work, large, centralized repositories tend to be more useful, but these are still not available in veterinary medicine. Also, these problems are further complicated by the fact that the data collection methods in different veterinary clinics vary. As an illustration, the quality of imaging equipment in the veterinary clinic might not be the same, creating discrepancies in the work of AI models. In response to this, scholars of veterinary medicine recommend that global, standardized datasets, covering a variety of species and medical cases, be used [81, 82]. Technical Challenges. These variations arise due to the incorporation of various modalities in both veterinary and human medicine, but the challenges may be of different magnitudes and resource supply. The first difficulty is the need to integrate data of different formats and resolutions, e.g., integrating high-resolution imaging information with low-resolution textual or audio input. The article [74] has pointed out the problems in the scale and features of modality peculiar to human diagnosis systems, as the lack of compatibility between them complicates data integration. These difficulties are further increased in the veterinary field as resources in a clinic are normally limited and, in most cases, do not have access to the same amount of computational infrastructure found in a human medical facility [50]. Research on veterinary medicine has tried to tackle these problems by creating lighter models that are not too large to maintain accuracy without overwhelming computational devices. These strategies are based on the progress in human medicine whereby solutions are implemented in resource-constrained health care settings [84]. Adoption Barriers The cost, lack of accessibility, and distrust of AI systems are the barriers to adoption in veterinary medicine. Implementation costs are also very high and can include equipment, software licenses, and training, and this is a major challenge, especially to the smaller clinics in low-resource environments. Such financial limitations are exaggerated in veterinary practice than in human care, where institutional funding or government subsidies tend to cover start-up expenses [79]. Also, in most cases, veterinarians are sceptical about the reliability and interpretability of AI-generated insights. The literature points out that one of the most critical issues in the use of AI models in veterinary and human medicine is the lack of insight into its internal processes, which discourages its application in the clinic. This problem is further complicated by the fact that practitioners in veterinary facilities lack technical skills. The gap between AI development and clinical practice is due to many veterinarians not having the required training to work with AI tools. To deal with this, special educational programs and cooperation between authors of AI and veterinary teachers are needed. These are reflective of other comparable efforts in human medicine, whereby AI literacy programmes are progressively being incorporated into medical education [82, 84]. Also, the issue of computational feasibility is becoming a concern especially in clinics that lack adequate infrastructure or connectivity. There are also solutions emerging to overcome this obstacle with AI methods that do not consume many resources. By facilitating decentralized model training on a multi-clinic basis, with no central data exchange, federated learning would limit privacy risks and hardware needs. Edge computing makes inference and low-level processing capabilities available on local devices by removing the necessity of a full-time connection to high-bandwidth or associated cloud computing. In addition, compression methods in models, including pruning and quantization, decrease the computational cost of deep-learning models without decreasing diagnostic quality [85- 87].

Ethical and Legal Concerns

The ethical and legal issues on multimodal artificial intelligence are slightly different in veterinary and human medicine but have similar themes. These issues are usually concerned with the secure disposal, storage, and utilization of sensitive data, including medical histories and imaging scans supplied by clients. Client confidentiality is crucial and must be upheld as well as data protection laws met so that ethical standards can be considered. In medical practice of human beings, the law on patient data, including the General Data Protection Regulation or the Health Insurance Portability and Accountability Act, is better established. However, there is often not an equivalent regulation in veterinary practices and, as a result, there are discrepancies in dealing with privacy and ownership of the data. Besides, the legal acknowledgement of animals differs considerably depending on jurisdictions, which affects the practice of veterinary medicine. Though animals are regarded as sentient beings in some countries, and hold a welfare status, others regard them as property and this makes their owners and veterinarians less responsible. Another ethical issue that is very serious in both disciplines is bias in AI models. When training is used in veterinary applications, species diversity in training data may cause biased or incorrect predictions due to under-represented population of animals. Equally, imbalanced data in human medical care can be used to marginalize certain demographic groups, thus highlighting the need to make datasets diverse in representing fairness and accuracy in all applications. Moreover, the principles of AI ethics in veterinary medicine should consider that animals are unable to give informed consent. In contrast to human patients whose decision-making on ways to proceed with their medical care is safeguarded by law and ethics, animals have to depend on their owners or caretakers to make medical decisions. This brings up the issue of possible contradiction of interest particularly in situations whereby financial constraints affect the decision to provide veterinary care. In certain states, AI-based diagnostics can be perceived as one of the cost-saving decisions; in others, they might be perceived as one of the unwarranted expenditures because of the legal definitions of animal ownership and care requirements. To solve these ethical challenges, legal frameworks must be clear, data-sharing agreements should be standardized, and efforts made to foster transparency and equity in the development of the AI models in the field of veterinary and human-focused technology. It is also mandatory to coordinate international collaboration to provide the harmonization of the regulations, making sure that the AIs-based veterinary diagnostics are in compliance with the principles of animal welfare and the changing legal norms worldwide. Creation of Comprehensive Data. The establishment of multimodal data sets of large scale dedicated to medicine in veterinary care is critical to enhance AI performance and provide the training of strong models. At the moment, numerous veterinary data do not have species-specific diversity and multimodal annotations, which impedes the extrapolation of AI models to other animal species. An urgent requirement is needed in the annotated datasets encompassing imaging, audio, text, and behavioral data to more reflect real-world veterinary clinical situations. Massive datasets, including electronic health records and imaging depositories, have played a crucial role in human medicine in driving the development of multimodal AI. The centralized datasets can be used as a standard of veterinary applications, where collection of data is frequently divided into single clinics. Interactions between educational establishments and veterinary clinics are essential to the aggregation of various and high-quality data. The suggested data-sharing contracts in the veterinary setting aim to eliminate the obstacles associated with the data fragmentation and privacy issues, respectively, and reflect the collaborative approach in human healthcare. **Improvements in Artificial Intelligence Methods**

Better methods of fusing multimodal data are a major research area in veterinary and human medicine, though the difficulties encountered in this area vary. As an example, although existing approaches including early, late and hybrid fusion have been effective in veterinary medicine, new approaches are needed to enable a smooth and real-time combination of various modalities. Early fusion models in veterinary practice have been demonstrated to be used to combine imaging with vital signs in veterinary diagnostics, with high accuracy in automated diagnostic assistance. These veterinary applications are based on techniques developed in human medicine including alignment algorithms, perception algorithms, which have shown higher accuracy because they combine the data inputs of various modalities. [82].

Explainable AI (XAI) has been identified as a priority in veterinary and human medicine, and the said objective is to address the current perception of AI systems as black boxes. The issue is particularly acute in the field of veterinary practice, the level of which is rather high in terms of distrust in the application of AI. Saliency maps, decision trees, SHAP values, and LIME are key strategies to improve the interpretability of a model by providing both visual and quantitative information about the decision making of a model. As an example, SHAP values can be used to explain the relative significance of audio and imaging features in the diagnosis of respiratory conditions. Research studies have proven the importance of such tools in helping veterinary professionals understand and have trust in AI outputs. [75].

Artificial intelligence, in pathogens diagnostic techniques

Artificial intelligence (AI) is extensively used to combat pathogens (parasites, bacteria, and viruses) in both human and animal health by enhancing diagnostics, accelerating drug discovery, and improving disease surveillance.

Artificial Intelligence (AI), as an interdisciplinary frontier dedicated to the emulation and expansion of human cognitive functions, pathogens, and the escalating crisis of antimicrobial resistance, AI serves as more than a computational tool; it acts as a critical catalyst for clinical innovation. By leveraging advanced modalities in data mining and predictive analytics, AI-driven frameworks are significantly refining diagnostic precision while empowering practitioners with proactive, real-time epidemiological surveillance.

This exploration categorizes the deployment of AI across diverse diagnostic environments, highlighting seminal advancements in high-throughput sequencing (HTS) for genomic tracking, the automated interpretation of radiological imagery, and the optimization of intelligent point-of-care testing (POCT) devices. However, the path toward seamless clinical translation remains a nuanced journey. A critical assessment reveals persistent barriers to widespread adoption, including the scarcity of high-fidelity longitudinal data, inherent limitations in algorithmic generalizability across diverse global populations, and the prevailing economic and regulatory hurdles. By consolidating recent empirical findings and cataloging pivotal data repositories, AI guides researchers through the lifecycle of integration from initial model architecture to the standardized validation of pathogen diagnosis.

The potential of AI as a research tool in the diagnosis of pathogens is tremendous. In this context, Parasites and Vectors has just launched the new article collection Artificial intelligence, parasites and parasitic diseases. This collection is dedicated to articles reporting the use of AI in parasitology research. I particularly welcome articles dealing with parasitic disease diagnosis, parasite and vector identification, as well as those reporting the prospection of drugs and vaccine candidates.

If you have a review proposal, please do not hesitate to contact me. This collection will serve as a platform for authors to publish their research that employs AI and deep learning methods for solving research questions in the fields of parasitology and tropical medicine.

Conclusion

Artificial intelligence is at the forefront of crucial issues in the world in terms of global health issues in many different areas. In antimicrobial resistance (AMR), AI can be used as a revolutionary tool because it can analyse large data volumes to determine drug targets, design novel antibiotics, and optimize antibiotic use to accelerate the creation of the necessary instruments in the fight against AMR. Regarding oncology, the contributions of AI in helping to improve the diagnostic techniques, treatment approaches, and drug development have potential in improving the outcomes of patients. The interdependence between zoonotic and reemerging diseases demands new methods, and AI in the field of health and veterinary studies is one of the main tools that can identify diseases early, contain them, and lead to the correct diagnosis, which was proven during the recent COVID-19 pandemic. AI has been found useful in genomic studies with applications in variant calling and annotation, and in personalized medicine strategies. Lastly, the impact of AI in drug discovery and development is characterized by streamlining the process, discovering new drug targets, and streamlining clinical trial designs, which gives way to more effective and efficient therapeutic solutions. The adoption of AI in all these areas of healthcare is an indicator of its disruptive capabilities in transforming the face of medical research and practice. Veterinary diagnostic imaging datasets have several challenges associated with them. To start with, the inconsistency of species and breeds occurring during the practice of veterinary medicine, and the typical heavy caseload of patients that leads to the accumulation of a vast amount of data, can be overwhelming. Second, the lack of typical cases of emerging diseases to train the algorithm is a significant disadvantage, and thus, some diagnoses cannot allowed to be provided unless one takes them into account in advance. Veterinary databases like VMDB should be embraced by the veterinary staff and professionals as it would give a platform on how to train AI algorithms on veterinary practice and research, based on availed data structure and images. Although there is increasing interest and effort to incorporate AI to be utilised in the human field of radiology in medicine, significant knowledge gap stems in successful full application of the same in veterinary medicine. Probably an amalgamation of new and old methodologies in the form of a hybrid solution would help in making a more acceptable shift towards newer, more powerful and useful predictive systems that would be easily adopted both by the practitioner and researchers. Multimodal AI is an innovative development in veterinary diagnostics that gives the opportunity to combine and process various types of data, such as imaging, behavioral video recordings, textual reports, and audio recordings, to support diagnostic aid in a holistic, efficient and accurate manner. Although the state of the art shows a large potential, the scaling to a wide use is still hindered by technical issues, including data integration and computational scalability, practical issues, including cost and lack of expertise, and ethical considerations, including data privacy and bias. To make the most of multimodal AI, a call to action should be issued, focusing on the necessity of the collaboration of researchers, veterinary doctors, AI developers, and policymakers. Creating high-quality large-scale data sets, enhancing the AI transparency by means of explainable AI, and incorporating easy-to-use tools into the clinical practice processes are the steps to be made. Moving forward, the future is obvious: multimodal AI can be turned into a regular part of veterinary practice, allowing clinicians to make data-based and accurate decisions and lighten the mental load, thus improving patient outcomes. Multimodal AI has the potential to redefine veterinary medicine by fixing the existing obstacles, improving diagnostic opportunities and enhancing animal care quality across the globe. The present review has recognized some of the limitations such as the fact that high-quality studies that specifically address veterinary multimodal AI were limited, and the scope of databases and keywords utilized might have created bias. Irrespective of these shortcomings, the systematic approach to the analysis used in this review will guarantee an extensive and methodological examination of the role of multimodal AI in veterinary diagnosis, which will be of great help to the research community, practitioners, and policymakers.

References

1. 1.Kour S, Agrawal R, Sharma N, Tikoo A, Pande N, Sawhney A. Artificial intelligence and its application in animal disease diagnosis. J Anim Res. (2022) 12:1. 10.30954/2277-940X.01.2022.1
2. 2.Kumar E. Artificial Intelligence. Delhi: IK International Pvt Ltd. (2013).
3. 3.Bohr A, Memarzadeh K. The Rise of Artificial Intelligence in Healthcare Applications. In Artificial Intelligence in Healthcare. Cambridge, MA: Academic Press; (2020). p. 25–60.
4. 4.Bouhali O, Bensmail H, Sheharyar A, David F, Johnson JP. A review of radiomics and artificial intelligence and their application in veterinary diagnostic imaging. Vet Sci. (2022) 9:620. 10.3390/vetsci9110620
5. 5.Ezanno P, Picault S, Beaunée G, Bailly X, Muñoz F, Duboz R, et al. Research perspectives on animal health in the era of artificial intelligence. Vet Res. (2021) 52:1–5. 10.1186/s13567-021-00902-4
6. 6.Reagan KL, Reagan BA, Gilor C. Machine learning algorithm as a diagnostic tool for hypoadrenocorticism in dogs. Domest Anim Endocrinol. (2020) 72:106396. 10.1016/j.domaniend.2019.106396
7. 7.Wanamaker MW, Vernau KM, Taylor SL, Cissell DD, Abdelhafez YG, Zwingenberger AL. Classification of neoplastic and inflammatory brain

disease using MRI texture analysis in 119 dogs. Vet Radiol Ultras. (2021) 62:445–54. 10.1111/vru.12962

8. 8.Agrebi S, Larbi A. Use of Artificial Intelligence in Infectious Diseases. In Artificial Intelligence in Precision Health. Cambridge, MA: Academic Press; (2020). p. 415–38.
9. 9.Guo W, Lv C, Guo M, Zhao Q, Yin X, Zhang L. Innovative applications of artificial intelligence in zoonotic disease management. Sci One Health. (2023) 2023:100045. 10.1016/j.soh.2023.100045
10. 10.Sebastian AM, Peter D. Artificial intelligence in cancer research: trends, challenges and future directions. Life. (2022) 12:1991. 10.3390/life12121991
11. 11.Xu B, Tan Z, Li K, Jiang T, Peng Y. Predicting the host of influenza viruses based on the word vector. PeerJ. (2017) 5:e3579. 10.7717/peerj.3579
12. 12.Babayan SA, Orton RJ, Streicker DG. Predicting reservoir hosts and arthropod vectors from evolutionary signatures in RNA virus genomes. Science. (2018) 362:577–80. 10.1126/science.aap9072
13. 13.Roy T, Sharma K, Dhall A, Patiyal S, Raghava GP. *In-silico* method for predicting infectious strains of Influenza A virus from its genome and protein sequences. bioRxiv. (2022). 10.1101/2022.03.20.485066
14. 14.Wang H, Cui W, Guo Y, Du Y, Zhou Y. Machine learning prediction of foodborne disease pathogens: algorithm development and validation study. JMIR Med Informat. (2021) 9:e24924. 10.2196/24924
15. 15.Guitian J, Snary EL, Arnold M, Chang Y. Applications of machine learning in animal and veterinary public health surveillance. Revue Scientifique et Technique. (2023) 42:230–41. 10.20506/rst.42.3366

16. 16.Ajayi T, Dara R, Poljak Z. Forecasting herd-level porcine epidemic diarrhea (PED) frequency trends in Ontario (Canada). Prev Vet Med. (2019) 164:15–22. 10.1016/j.prevetmed.2019.01.005
17. 17.Wardeh M, Blagrove MS, Sharkey KJ, Baylis M. Divide-and-conquer: machine-learning integrates mammalian and viral traits with network features to predict virus-mammal associations. Nat Commun. (2021) 12:3954. 10.1038/s41467-021-24085-w
18. 18.Munck N, Njage PM, Leekitcharoenphon P, Litrup E, Hald T. Application of whole-genome sequences and machine learning in source attribution of Salmonella Typhimurium. Risk Analysis. (2020) 40:1693–705. 10.1111/risa.13510
19. 19.Deneke C, Rentzsch R, Renard BY. PaPrBaG: a machine learning approach for the detection of novel pathogens from NGS data. Sci Rep. (2017) 7:39194. 10.1038/srep39194
20. 20.Sunuwar J, Azad RK. A machine learning framework to predict antibiotic resistance traits and yet unknown genes underlying resistance to specific antibiotics in bacterial strains. Brief Bioinformat. (2021) 22:bbab179. 10.1093/bib/bbab179
21. 21.Njage PM, Leekitcharoenphon P, Hald T. Improving hazard characterization in microbial risk assessment using next generation sequencing data and machine learning: predicting clinical outcomes in shigatoxigenic *Escherichia coli. Int J Food Microbiol.* (2019) 292:72–82. 10.1016/j.ijfoodmicro.2018.11.016
22. 22.Zuidema D, Kerns K, Sutovsky P. An exploration of current and perspective semen analysis and sperm selection for livestock artificial insemination. Animals. (2021) 11:3563. 10.3390/ani11123563

23. 23.Dhoble AS, Ryan KT, Lahiri P, Chen M, Pang X, Cardoso FC, et al. Cytometric fingerprinting and machine learning (CFML): a novel label-free, objective method for routine mastitis screening. Comput Electr Agri. (2019) 162:505–13. 10.1016/j.compag.2019.04.029
24. 24.Taneja M, Byabazaire J, Jalodia N, Davy A, Olariu C, Malone P. Machine learning based fog computing assisted data-driven approach for early lameness detection in dairy cattle. Comput Electr Agri. (2020) 171:105286. 10.1016/j.compag.2020.105286
25. 25.Keceli AS, Catal C, Kaya A, Tekinerdogan B. Development of a recurrent neural networks-based calving prediction model using activity and behavioral data. Comput Electr Agri. (2020) 170:105285. 10.1016/j.compag.2020.105285
26. 26.Murphy MD, O'Mahony MJ, Shalloo L, French P, Upton J. Comparison of modelling techniques for milk-production forecasting. J Dairy Sci. (2014) 97:3352–63. 10.3168/jds.2013-7451
27. 27.Rabaan AA, Alhumaid S, Mutair AA, Garout M, Abulhamayel Y, Halwani MA, et al. Application of artificial intelligence in combating high antimicrobial resistance rates. Antibiotics. (2022) 11:784. 10.3390/antibiotics11060784
28. 28.Pereira AI, Franco-Gonçalo P, Leite P, Ribeiro A, Alves-Pimenta MS, Colaço B, et al. Artificial intelligence in veterinary imaging: an overview. Vet Sci. (2023) 10:320. 10.3390/vetsci10050320
29. 29.Fraiwan MA, Abutarbush SM. Using artificial intelligence to predict survivability likelihood and need for surgery in horses presented with acute abdomen (colic). J Equine Vet Sci. (2020) 90:102973. 10.1016/j.jevs.2020.102973

30. 30.Anwar SM, Majid M, Qayyum A, Awais M, Alnowami M, Khan MK. Medical image analysis using convolutional neural networks: a review. J Med Syst. (2018) 42:1–3. 10.1007/s10916-018-1088-1
31. 31.Esteva A, Robicquet A, Ramsundar B, Kuleshov V, DePristo M, Chou K, et al. A guide to deep learning in healthcare. Nat Med. (2019) 25:24–9. 10.1038/s41591-018-0316-z
32. 32.Hur B, Hardefeldt LY, Verspoor KM, Baldwin T, Gilkerson JR. Evaluating the dose, indication and agreement with guidelines of antimicrobial use in companion animal practice with natural language processing. JAC Antimicrob Resist. (2022) 4:dlab194. 10.1093/jacamr/dlab194
33. 33.Ali T, Ahmed S, Aslam M. Artificial intelligence for antimicrobial resistance prediction: challenges and opportunities towards practical implementation. Antibiotics. (2023) 12:523. 10.3390/antibiotics12030523
34. 34.Lv J, Deng S, Zhang L. A review of artificial intelligence applications for antimicrobial resistance. Biosaf Health. (2021) 3:22–31. 10.1016/j.bsheal.2020.08.003
35. 35.Masud RI, Fahim NA, Rana ML, Islam MS, Tanvir M. Artificial intelligence, a powerful tool to combat antimicrobial resistance: an update. J Adv Biotechnol Exp Ther. (2023) 6:711–27. 10.5455/jabet.2023.d161
36. 36.Zhou S, Liu B, Han Y, Wang Y, Chen L, Wu Z, et al. ZOVER: the database of zoonotic and vector-borne viruses. Nucleic Acids Res. (2022) 50:D943–9. 10.1093/nar/gkab862
37. 37.Zhou X, Lee EWJ, Wang X, Lin L, Xuan Z, Wu D, et al. Infectious diseases prevention and control using an integrated health big data system in China. BMC Infect Dis. (2022) 22:344. 10.1186/s12879-022-07316-3

38. 38.Galiez C, Siebert M, Enault F, Vincent J, Söding J. WIsH: who is the host? Predicting prokaryotic hosts from metagenomic phage contigs. Bioinformatics. (2017) 33:3113–4. 10.1093/bioinformatics/btx383
39. 39.Liang G, Fan W, Luo H, Zhu X. The emerging roles of artificial intelligence in cancer drug development and precision therapy. Biomed Pharmacother. (2020) 128:110255. 10.1016/j.biopha.2020.110255
40. 40.Luchini C, Pea A, Scarpa A. Artificial intelligence in oncology: current applications and future perspectives. Br J Cancer. (2022) 126:4–9. 10.1038/s41416-021-01633-1
41. 41.Gregory AV, Kline TL. The Role of Artificial Intelligence in Automated Data Analysis of the Kidney. In Advanced Clinical MRI of the Kidney: Methods and Protocols. Cham: Springer International Publishing; (2023). p. 419–36.
42. 42.Mikhael PG, Wohlwend J, Yala A, Karstens L, Xiang J, Takigami AK, et al. Sybil: a validated deep learning model to predict future lung cancer risk from a single low-dose chest computed tomography. J Clin Oncol. (2023) 41:2191–200. 10.1200/JCO.22.01345
43. 43.Dias R, Torkamani A. Artificial intelligence in clinical and genomic diagnostics. Genome Med. (2019) 11:1–2. 10.1186/s13073-019-0689-8
44. 44.Sharma A, Virmani T, Pathak V, Sharma A, Pathak K, Kumar G, et al. Artificial intelligence-based data-driven strategy to accelerate research, development, and clinical trials of COVID vaccine. BioMed Res Int. (2022) 2022:7205241. 10.1155/2022/7205241
45. 45.Mazzocco G, Niemiec I, Myronov A, Skoczylas P, Kaczmarczyk J, Sanecka-Duin A, et al. AI aided design of epitope-based vaccine for the

induction of cellular immune responses against SARS-CoV-2. Front Genet. (2021) 12:602196. 10.3389/fgene.2021.602196

46. 46.Blanco-Gonzalez A, Cabezon A, Seco-Gonzalez A, Conde-Torres D, Antelo-Riveiro P, Pineiro A, et al. The role of ai in drug discovery: challenges, opportunities, and strategies. Pharmaceuticals. (2023) 16:891. 10.3390/ph16060891
47. C. Z. L. R. Soenksen, Y. Ma, and L. Boussioux, "Integrated multimodal artificial intelligence framework for healthcare applications," *NPJ Digit. Med.*, vol. 5, pp. 400–410, Jan. 2022. [Online]. Available: https://www.nature.com/articles/s41746-022-00689-4
48. D. Yitbarek and G. G. Dagnaw, "Application of advanced imaging modalities in veterinary medicine: A review," *Vet. Med., Res. Rep.*, vol. 14, pp. 117–130, May 2022. [Online]. Available: https://www.tandfonline.com/doi/pdf/10.2147/VMRR.S367040
49. M. Latif, M. S. Arshad, M. Usman, M. Mobashar, and Z. Wali, "AI advancements in veterinary diagnostics," *Continental Vet. J.*, vol. 24, no. 5, pp. 502–520, 2024. [Online]. Available: https://cvetj.com/upload/issue_paper/1722722224_CVJ-24-0502.pdf
50. M. Burić and M. Ivašić-Kos, "AI in veterinary ophthalmology: Canine eye diseases diagnosis with image segmentation and large language models," in *Proc. World Conf. Smart Trends Syst., Secur., Sustainability*, Jan. 2025, pp. 381–390. [Online]. Available: https://www.researchgate.net/publication/384689681

51. R. B. Appleby and P. S. Basran, “Artificial intelligence in diagnostic imaging,” *Adv. Small Animal Care*, vol. 5, no. 2, pp. 123–135, 2024. [Online]. Available: https://www.advancesinsmallanimalcare.com/article/S2666-450X

52. D. Leary and P. S. Basran, “The role of artificial intelligence in veterinary radiation oncology,” *Vet. Radiol. Ultrasound*, vol. 63, no. S1, pp. 903–912, Dec. 2022. [Online]. Available: https://onlinelibrary.wiley.com/doi/abs/10.1111/vru.13162

53. Y. Dhaliwal, H. Bi, and S. Neethirajan, “Bimodal data analysis for early detection of lameness in dairy cows using artificial intelligence,” in *Proc. SSRN AI Vet. Symp.*, Jan. 2024, pp. 1–8. [Online]. Available: https://papers.ssrn.com/sol3/Delivery.cfm?abstractid=4976718

54. T. A. Patil, “Optimizing animal rehabilitation and physical therapy: AI-driven evidence-based strategies for achieving optimal outcomes in veterinary practice,” *Revista Electrnica de Veterinaria*, vol. 12, no. 3, pp. 134–145, 2024. [Online]. Available: https://veterinaria.org/index.php/REDVET/article/download/528/307

55. M. Thißen, T. N. D. Tran, B. J. Schönbein, U. Trapp, B. E. Ratsch, B. Egner, R. Piat, and E. Hergenröther, “Enhancing canine musculoskeletal diagnoses: Leveraging synthetic image data for pre-training AI-models on visual documentations,” 2024, *arXiv:2409.08181*.

56. M. M. Amer, A. M. Amer, and K. M. El-Bayoumi, “Artificial intelligence: Current and future role in veterinary and public medicine,” *Egyptian J. Vet. Sci.*, pp. 1–12, Aug. 2024. [Online]. Available: https://journals.ekb.eg/article_376137_97783c106d904e3d2196221315eb73c0.pdf and https://ejvs.journals.ekb.eg/article_376137.html

57. J. M. Perez, "The need of a proof-of-concept in artificial intelligence-based health-related quality of life instruments in veterinary medicine," *Imag. Radiat. Res.*, vol. 13, no. 4, pp. 243–260, 2023. [Online]. Available: https://systems.enpress-publisher.com/index.php/IRR/article/view/2201
58. T. Baltrusaitis, C. Ahuja, and L.-P. Morency, "Multimodal machine learning: A survey and taxonomy," *IEEE Trans. Pattern Anal. Mach. Intell.*, vol. 41, no. 2, pp. 423–443, Feb. 2019.
59. M. Tamošiūnas, O. Čiževskis, D. Viškere, M. Melderis, U. Rubins, and B. Cugmas, "Multimodal approach of optical coherence tomography and Raman spectroscopy can improve differentiating benign and malignant skin tumors in animal patients," *Cancers*, vol. 14, no. 12, p. 2820, Jun. 2022. [Online]. Available: https://www.mdpi.com/2072-6694/14/12/2820/pdf
60. J. W. J. L. Y. Zhang, Y. Yin, and X. Wang, "Improving medical diagnosis with multimodal learning: Methods, challenges, and applications," *IEEE J. Biomed. Health Informat.*, vol. 26, no. 5, pp. 2154–2167, Oct. 2022.
61. G. Currie and E. Rohren, "Intelligent imaging: Applications of machine learning and deep learning in radiology," *Vet. Radiol. Ultrasound*, vol. 63, no. S1, pp. 880–888, Dec. 2022. [Online]. Available: https://onlinelibrary.wiley.com/doi/abs/10.1111/vru.13144
62. J. Chen, N. Zhao, V. Copello, Y. Ru, K. L. Burnstein, and Y. Yang, "Accurate and early metastases diagnosis in live animals with multimodal X-ray and optical imaging," *Int. J. Radiat. Oncol.*, vol. 15, no. 3, pp. 215–230, 2023. [Online]. Available: https://www.sciencedirect.com/science/article/pii/S0360301622025871

63. L. Zhu, W. Mou, Y. Lai, J. Chen, S. Lin, L. Xu, J. Lin, Z. Guo, T. Yang, A. Lin, C. Qi, L. Gan, J. Zhang, and P. Luo, "Step into the era of large multimodal models: A pilot study on ChatGPT-4V's ability to interpret radiological images," *Int. J. Surg.*, vol. 90, pp. 1–15, Jan. 2024. [Online]. Available: https://journals.lww.com/international-journal-of-surgery/citation/9900/step_into_the_era_of_large_multimodal_models__a.1218.aspx

64. Zuraw and F. Aeffner, "Whole-slide imaging, tissue image analysis, and artificial intelligence in veterinary pathology: An updated introduction and review," *Vet. Pathol.*, vol. 59, no. 1, pp. 6–25, Jan. 2022. [Online]. Available: https://journals.sagepub.com/doi/pdf/10.1177/03009858211040484

65. C. Gomes, P. Tilley, and L. Coheur, "Evaluating asthma in equines with video recordings," in *Proc. EPIA Conf. Artif. Intell.* Cham, Switzerland : Springer, Nov. 2024, pp. 38–49.

66. C. T. R. W. Bello, O. Fatoki, and R. Hans, "Role of computer vision and deep learning algorithms in livestock behavioral recognition: A state-of-the-art review," *Edelweiss Appl. Sci. Technol.*, vol. 8, no. 6, pp. 6416–6430, Jan. 2024. [Online]. Available: https://www.researchgate.net/publication/386273963

67. Franco, M. Russo, M. Amboni, A. M. Ponsiglione, F. Di Filippo, M. Romano, F. Amato, and C. Ricciardi, "The role of deep learning and gait analysis in Parkinson's disease: A systematic review," *Sensors*, vol. 24, no. 18, p. 5957, Sep. 2024. [Online]. Available: https://www.mdpi.com/1424-8220/24/18/5957

68. N. S. J. M. Neary and G. Oikonomou, "Automated dairy cattle lameness detection utilizing the power of artificial intelligence: Current status quo and

future research opportunities," *Vet. J.*, vol. 210, pp. 45–60, Jan. 2024. [Online].
Available: https://www.sciencedirect.com/science/article/pii/S109002332400 0303

69. N. Luo, X. Zhong, L. Su, Z. Cheng, W. Ma, and P. Hao, "Artificial intelligence-assisted dermatology diagnosis: From unimodal to multimodal," *Comput. Biol. Med.*, vol. 150, pp. 122–130, Oct. 2023. [Online]. Available: https://www.sciencedirect.com/science/article/pii/S001048252300 8788
70. Q. Niu, K. Chen, M. Li, P. Feng, Z. Bi, L. K. Q. Yan, Y. Zhang, C. H. Yin, C. Fei, J. Liu, B. Peng, T. Wang, Y. Wang, S. Chen, and M. Liu, "From text to multimodality: Exploring the evolution and impact of large language models in medical practice," 2024, *arXiv:2410.01812*.
71. Z. Sun, M. Lin, Q. Zhu, Q. Xie, F. Wang, Z. Lu, and Y. Peng, "A scoping review on multimodal deep learning in biomedical images and texts," *J. Biomed. Informat.*, vol. 153, pp. 78–90, Oct. 2023. [Online]. Available: https://www.sciencedirect.com/science/article/pii/S153204642300 2034
72. M. N. Reza, M. R. Ali, M. A. Haque, H. Jin, H. Kyoung, Y. K. Choi, G. Kim, and S.-O. Chung, "A review of sound-based pig monitoring for enhanced precision production," *J. Animal Sci. Technol.*, vol. 67, no. 2, pp. 277–302, Mar. 2025. [Online]. Available: https://www.ejast.org/archive/view_article?pid=jast-2024-e113
73. O. Nwaiwu, M. Rahman, and F. Alam, "Ai-powered bioacoustic monitoring in livestock: A review of applications and challenges," *Frontiers Vet. Sci.*, vol. 11, Jan. 2024, Art. no. 1178452.

74. Y. Zhang, T. Xia, A. Saeed, and C. Mascolo, "RespLLM: Unifying audio and text with multimodal LLMs for generalized respiratory health prediction," 2024, *arXiv:2410.05361*.

75. S. R. Chetupalli, P. Krishnan, N. Sharma, A. Muguli, R. Kumar, V. Nanda, L. M. Pinto, P. K. Ghosh, and S. Ganapathy, "Multi-modal point-of-care diagnostics (don't short) for COVID-19 based on acoustics and symptoms," *IEEE J. Transl. Eng. Health Med.*, vol. 11, pp. 199–210, 2023. [Online]. Available: https://ieeexplore.ieee.org/abstract/document/10064207/

76. Megahed, R. Bommineni, M. Short, K. N. Galvão, and J. H. J. Bittar, "Using supervised machine learning algorithms to predict bovine leukemia virus seropositivity in dairy cattle in florida: A 10-year retrospective study," *Preventive Veterinary Med.*, vol. 235, Feb. 2025, Art. no. 106387. [Online]. Available: https://www.sciencedirect.com/science/article/pii/S016758772400 2733

77. S. Mehta and S. Choudhary, "Advancements in veterinary diagnostics: A CNN-SVM model for cattle disease identification," in *Proc. 4th Asian Conf. Innov. Technol. (ASIANCON)*, Aug. 2024, pp. 1–5. [Online]. Available: https://ieeexplore.ieee.org/abstract/document/10838102/

78. S. Lububu and M. Twum-Darko, "Comparative analysis of machine learning algorithms to improve the diagnosis of African swine fever," *Int. J. Bus. Ecosyst. Strategy*, vol. 6, no. 5, pp. 121–137, Dec. 2024. [Online]. Available: https://www.bussecon.com/ojs/index.php/ijbes/article/view/646

79. Tanaka and K. Shimada. (2024). *Multi-Sensor Fusion to Enhance Accuracy in Veterinary Diagnostics*. [Online]. Available: https://www.preprints.org/frontend/manuscript/5d0f72/download

80. D. Brezov, H. Hristov, D. Dimov, and K. Alexiev, "Predicting the rectal temperature of dairy cows using infrared thermography and multimodal machine learning," *Appl. Sci.*, vol. 13, no. 20, p. 11416, Oct. 2023. [Online]. Available: https://www.mdpi.com/2076-3417/13/20/11416
81. N. J. Acheme. (2024). *Deep Convolutional Neural Networks for Dog Heart Image Classification.* ResearchGate. [Online]. Available: https://www.researchgate.net/publication/385943432_Deep_Convolutional_Neural_Networks_for_Dog_Heart_Image_Classification
82. R. T. A. S. Mandour and X. Zhao. (2024). *Remote Vital Sensing in Clinical Veterinary Medicine: Advances, Accomplishments, Challenges, and Future Perspectives.* [Online]. Available: https://www.preprints.org/frontend/manuscript/dcf90c06f27421e814fce467feab98d4/download_pub
83. Meoded. (2024). *Dog Cardiomegaly Diagnosis With Explainable Keypoint Detection.* [Online]. Available: https://www.researchgate.net/publication/386425209
84. Dave T, Athaluri SA, Singh S. ChatGPT in medicine: an overview of its applications, advantages, limitations, future prospects, and ethical con siderations. Front Artif Intell. 2023;6:1169595.
85. Salvagno M, Taccone FS, Gerli AG. Can artificial intelligence help for scientific writing? Crit Care. 2023;27:75.
86. Huang J, Tan M. The role of ChatGPT in scientific communication: writing better scientific review articles. Am J Cancer Res. 2023;13:1148–54.
87. Sinclair BJ. Letting ChatGPT do your science is fraudulent (and a bad idea), but AI-generated text can enhance inclusiveness in publishing. Curr Res Insect Sci. 2023;3:100057.

88. Mojadeddi ZM, Rosenberg J. The impact of AI and ChatGPT on research reporting. N Z Med J. 2023;136:60–4.

89. Stokel-Walker C. ChatGPT listed as author on research papers: many scientists disapprove. Nature. 2023;613:620–1.

90. Da Re D, Van Bortel W, Reuss F, Müller R, Boyer S, Montarsi F, et al. DynamAedes: a unified modelling framework for invasive Aedes mosqui toes. Parasit Vectors. 2022;15:414.

91. Poché DM, Wagner D, Green K, Smith Z, Hawthorne N, Tseveenjav B, et al. Development of a low-dose fipronil deer feed: evaluation of efficacy against two medically important tick species parasitizing white-tailed deer (Odocoileus virginianus) under pen conditions. Parasit Vectors. 2023;16:94.

Chapter (5): Between Seed and Data: Agriculture as a Knowledge System in the Age of Artificial Intelligence

Sajida Saeed Aziz
Department of Horticulture, College of Agricultural Engineering Sciences, University of Sulaimani, Iraq
E-mail: Sajida.aziz@univsul.edu.iq

Abstract:

Agriculture has a great effect on the economic sector. Artificial intelligence is being used in the agricultural sector to automate many activities and minimize risks and offer a relatively user-friendly farming system. However, there is a revolution in the agricultural sector whereby artificial intelligence (AI) is being utilized to solve the key problems that affect the agricultural sector in developing countries. This chapter discusses a focus on the effect and the potential benefits of AI in the developing countries. A lack of infrastructure and presence of high costs of investment and capacity building have been cited to be the challenges facing the widespread use of AI in agriculture in developing nations. It requires governments, policymakers, researchers, and the providers of technologies to cooperate in order to make the use of AI technologies in agriculture sustainable and inclusive. The main aim of the study is a critical analysis of the situation in the agricultural sector of developing countries, as well as the detection and removal of the difficulties of farmers who have been exposed to AI technologies. However, there is a revolution unfolding in the agricultural sector, which applies the artificial intelligence (AI) to address some of the biggest problems that the agricultural sector faces in developing countries.

Key words: Seed, Data, Agriculture, a Knowledge System, Artificial Intelligence

Introduction

The shift Between Seed and Data: Agriculture as a Knowledge System in the Age of Artificial Intelligence outlines the paradigmmatic change in which farming is no longer based on the traditional, intuitive wisdom but is an information-driven business with an information-centric approach. This shift transforms the nature of agricultural knowledge production, storage, and use, and instead of the generational knowledge system that relied on the physical seed, such as a dynamic system that worked through artificial intelligence (AI) and large datasets. The rising demand of agricultural products is ever escalating with the high rate of population growth all over the globe. Crop yield production is a requirement that is set to increase its quantity by 2050 meaning that the figure will escalate by about 2.4 each year. Nevertheless, an annual loss of crop yields of approximately 20% - 40% is caused by the disease of plants and insect pests, which is aggravated by the lack of the water resource needed to cultivate crops. The use of plant disease identification has become an important issue in the recent years. Tomatoes, potatoes, bell peppers are among the other fruits and vegetables that are used. It is important to detect diseases in the leaves of these crops in time since the plant leaves are the first method of identifying the infection on the leaves since most of the disease symptoms can be observed on the leaves.[1-4].

An automated disease detection and classification system based on images is a necessary measure to keep the ecosystem healthy and to guarantee food production. Two-way Agricultural techniques, in particular, image processing and deep learning, alleviate the problem of machine-learning and offer promising solutions to disease detection and control, which can transform the agricultural practice. Such technologies use complex image recognition to get rid of the subjective judgments and provide more accurate outcomes. Early detection of crop diseases is essential in the implementation of required interventions and the avoidance of massive losses in large-scale agricultural whatsoever. [5].

Machine learning has drawbacks which can be overcome by the adaptive transfer learning method as a deep learning method. It performs better on the target task using fewer data and computational resources than the inefficiencies of training a model form scratch. This saves the need to have massive pieces of computational resource and improves the generalization capability of the model to a wide range of datasets. In this regard, the transfer learning was adopted in our study, as it possesses considerable benefits. The deep learning technique of transfer learning was used to detect plant disease in this paper [6]. Various sophisticated methods have been published to not only solve the problem of plant diseases but also to conserve water irrigation and avoid pest propagation among other issues. Such methods must be incorporated in the farm machines like drones and robotics, especially to manage farm activities in the farming sector. Drones or Unmanned Aerial Vehicles (UAVs) are an emerging technology that promises to be effective in remote farming. UAVs are remotely operated and wirelessly linked with the WSN technology in order to fly around the extensive field of any agricultural applications like identification of plant diseases and prevention of pests attacks. The agricultural field has over 250 models all over the world. [7-9].

Depending on the intended agricultural application, UAVs can be fitted with RGB, hyperspectral, or multispectral cameras to take aerial shots of the planted fields, to build large datasets to learn deep-learning algorithms to predict plant diseases, pesticides spraying, and plant development [10-12]. Most of the UAV models, each with its own set of characteristics and properties, can be utilized in two main agricultural tasks: (1) the acquisition of real-time aerial images to monitor crop yields and the vegetation landscape; (2) the spraying of liquids: fertilizers, toxic substances, water, etc., to treat diseases, eliminate pests and water, respectively. However, it is not easy to design and implement a single UAV model that would perform both functions. The UAV design requires endurance, weight, speed and cost to be considered and evaluated beforehand, depending on the nature of farming operation and requirements. Some of the disadvantages associated with UAVs include high prices, poor battery life, and range of communication, which in some cases make them inadequate to complete all the necessary farming operations. Robots may on the other hand substitute human beings in order to have more efficient and intelligent control of crops. Robotics performs multistep operations in agricultural applications that are time and labor and cost-saving in multiple applications. They will be able to know the best time to irrigate the land, when to plant the seedlings, weeding, when to apply fertilisers, application of pesticides to control pests and they will also know when a plant is diseased. Nevertheless, ground robots are limited to barriers, rough surfaces, route planning, slower velocity, and less hardware is required to carry out several farm duties [13-19]. Rather than extending the previous studies, the proposed paper will combine image-processing and deep-learning algorithms to detect plant diseases with a high level of protection. Both of the methods are implemented on the suggested hardware pilot intended to treat plant diseases. The hardware pilot is fixed to a pivot which is centrally located, thus eliminating the use of UAVs and robotics, and limiting their shortcomings. Introducing image-processing and deep-learning functions into a central pivot will be an artificial-intelligence-Internet-of-Things (AI-IoT) smart agriculture pivot. [13].

1. The proposed AI-IoT smart agricultural pivot will be used instead of UAVs and will be capable of executing all the primary agricultural missions that would be

carried out by separate UAV models, hence covering a wide range of areas of a farm field. The pivot is on a height close to the crops and it allows easy access to tall vegetation and application of liquid treatment on plant diseases, pest control and irrigation. It has a long arm that gives it a wide reach of farmlands, which most aerial harvesting operations were traditionally performed by UAVs; it can conduct most of the functions of detecting plant diseases without the high cost and power-intensive multi-spectral or hybrid-spectral cameras. The pivot can also perform all the farming tasks in a low height of about 23 m of about 2-3 m above the ground hence avoids the constraints associated with robotics [13]. In Plant disease detection, a convolutional neural network model has been constructed to detect leaf diseases, using visual data, which improves the accuracy, generality and general training efficiency. The experimental outcomes reveal that the proposed convolutional neural network-based model is superior to the earlier models, with a classification accuracy of 99.23 0 -percent [20]. Sun et al. proposed FL-EfficientNet (Focal loss EfficientNet) to identify multiple categories of plant diseases. On the public New Plant Diseases Dataset (NPDD), FL-EfficientNet performed comparatively to ResNet50, DenseNet169 and EfficientNet with an accuracy rate of 99.72 percent in detecting ten diseases in five crop types. The greatest classification accuracy of 99.3 was found with this model [21,'22].a.

Jerome et al. [23] studied a deep-learning guided evaluation-driven convolutional neural network (A -CNN) in classifying healthy and diseased plant leaves. This method was implemented on the detection of pervasive plant maladies, and 92% of accuracy on classification of the disease is attained. T.-Y. Lee et al. proposed a high-efficiency CNN architecture with specific designs that were tailored to detecting potato diseases in reference [24]. The authors created a training set based on image-processing. They were using Adam optimizer and the loss of cross-entropy to train a model and a Softmax layer as a final decision unit. The obtained CNN achieved an accuracy of 99.53. [23, 24].

A central pivot has been recently added as a mechanical and electrical pilot device as an effective irrigation device. It was designed in a way that guaranteed equal allocation of water in the farmland. The pivot system provided the same amount of water to overall space and therefore, it became especially efficient when it comes to watering circular farms. The sprinklers were under control by a control unit (CU) which ensured constant water volume and flow rate. In 2028, A. Shilpa et al. used the IoT technology to create a more traditional pivot system, which they called center pivot irrigation (CPI), a type of self-driven irrigation equipment that calculates the percentage of irrigation applied in each section of a divided area of a round farm and considers the differences in topography and soil types. The CPI could be applied to areas that had irregular topography and also there were no restrictions to crops. Circular irrigation has the capability of covering about 76 percent of the field; a further extension of the irrigated acreage is achieved with the addition of an end gun and consequently the area that lacks irrigation is minimized. The Self-Propelled CPI System is an IoT-based system to keep track of temperature and soil moisture readings, which are then sent through an internet gateway to a database to be stored, analyzed and smart decisions made on the need to apply irrigation. To irrigate water and apply fertilizer and not to detect any disease in plants, Badreldeen et al. enriched the central pivot system (CPS) with rain and humidity sensors, which enhanced the management of irrigation and saved water resources. [25-29]

To monitor crop irrigation, Matilla et al. proposed centre-pivot irrigation monitoring systems using IoT technology and two low-cost communication technologies (LoRaWAN and GPS) to report the pivot status and identify operation changes such as unwanted stoppage, and this way allow the farmers to monitor the pivot sensors. Both wireless technologies overcame the long-range communication obstacles that were foreseen in the expansive field environment and in remote locations, hence a successful irrigation monitoring system. The suggested system will give the detailed electrical description of the interfacing between LoRaWAN and GPS and the control unit of the pivot, as well as a case study will prove the idea. The metrics of packet loss and received packets were considered to determine the consequence on the running of the system. [30].

O. Debauche et al. suggested a multi-agent system (MAS) built on the Kubernetes and Docker infrastructure to scale the AI algorithms on heterogeneous clusters in the edge to compute the water usage and record videos of plant illnesses close to the pivot and, therefore, reduce the processing time to implement the rule-based decisions. The functionality of the proposed architecture was tested by various AI algorithms on water requirement data and a plant villages database in order to validate their functionality. However, the issue of pest detection was not discussed in these works [31, 32]. The given work will involve the AI-IoT system structure and hardware prototype. The AI-IoT agricultural pivot system of smart potato plants with plant diseases detection and treatment is shown in Fig. 1.

Fig. 1

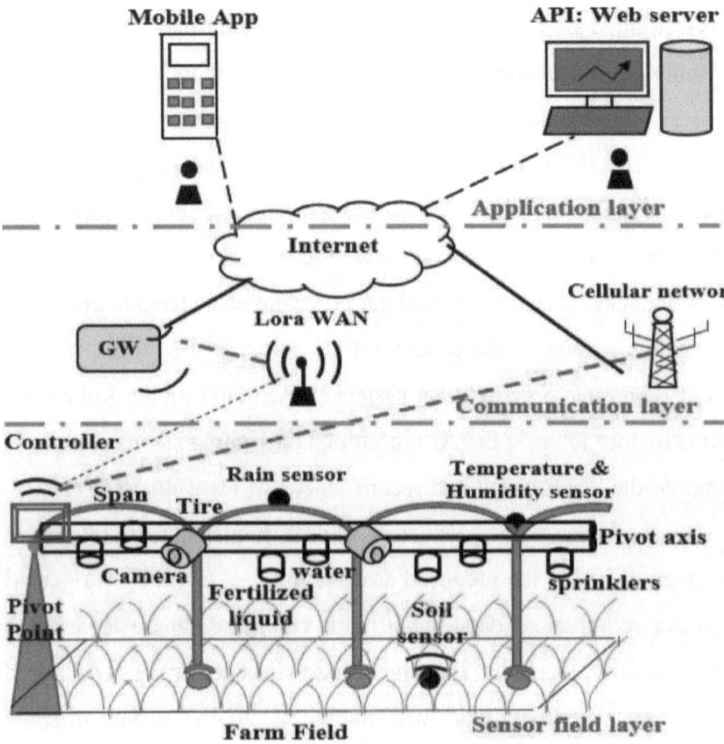

The proposed AI-IoT system architecture

Instead of the aforementioned ones, the proposed hardware pilot will be your hardware to be implemented on the central pivot directly as a part of the sensor field layer shown in Figure 1 to detect and treat plant diseases. As per the suggested system architecture, the hardware pilot is made up of three main levels of dependencies which include physical level, communication level, and network level [2832]. Communication level The communication level is the communication unit of the proposed architecture that will have the Raspberry Pi controller as its role. In this tier, the Raspberry Pi will be fitted with communication technologies such as Ethernet, Bluetooth, GSM/GPRS module, and Wi-Fi that will allow it interact remotely with the application layer via the internet. All the communication technologies were chosen according to their communication range. The suggested hardware pilot is constructed on the basis of the SIM800L GSM/GPRS module, which is one of a few wireless technologies that are broadly used to provide long-range communication. It uses SIM800L GSM/GPRS module to send the categorized pictures to the mobile app at the application layer through the available cellular network towers [33].

Data augmentation

Images captured will have to be re-processed to achieve very precise and clear features of images and then undergo plant disease classification. The process of augmentation has the power to equip the ResNet50 classifier model with the capability to detect incoming image features very accurately and with the lowest loss to detect plants. The processing techniques that are used to enhance the images include rescaling, shear range, zoom range, horizontal flip, and resizing. Rescaling brings pixel values in the range 0 255 to 0 1 by dividing each pixel value by 255.5, and in this way, the convergence of classifier model becomes faster when training. As an illustration, when the initial value of a captured picture is [120, 105], then the rescaled value will be [0.471, 0.412]. A shearing is a change which is a replacement of every point by a specific direction, changing the image skewly. The shear range of 0.2 means that the image will be sheared randomly by an angle up to 0.2 radians (almost 11 degrees), which adds variation, which makes the model resistant to such transformations in practice. It is a random zoom of a range of about -20 20 per cent of the image size, allowing the learning model to identify objects with a change in size, so that when the camera is zoomed in, it appears like it is closer, and when zoomed out, it seems like it is further away. The histogram of the original image pixels and that of the sheared image pixels (with a range of 0.2) are depicted in Figure 2 and Figure 3 respectively, and this indicates its effects on training data set in relation to original image pixels.

Fig. 2

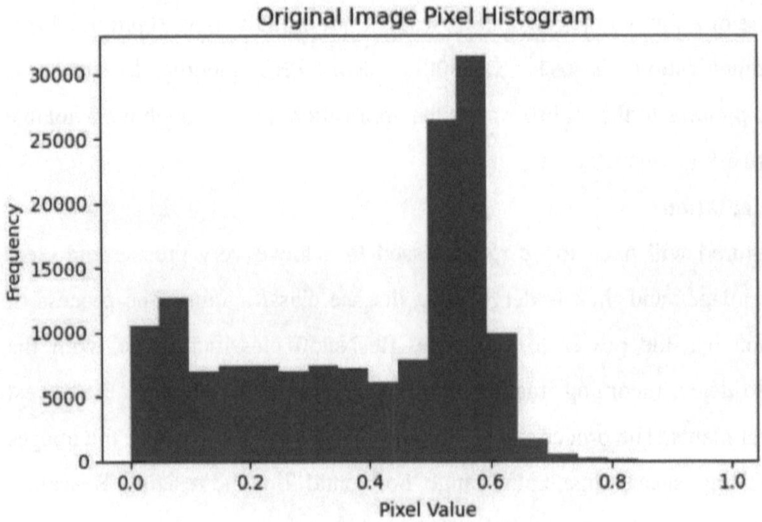

Original image pixel histogram

Fig. 3

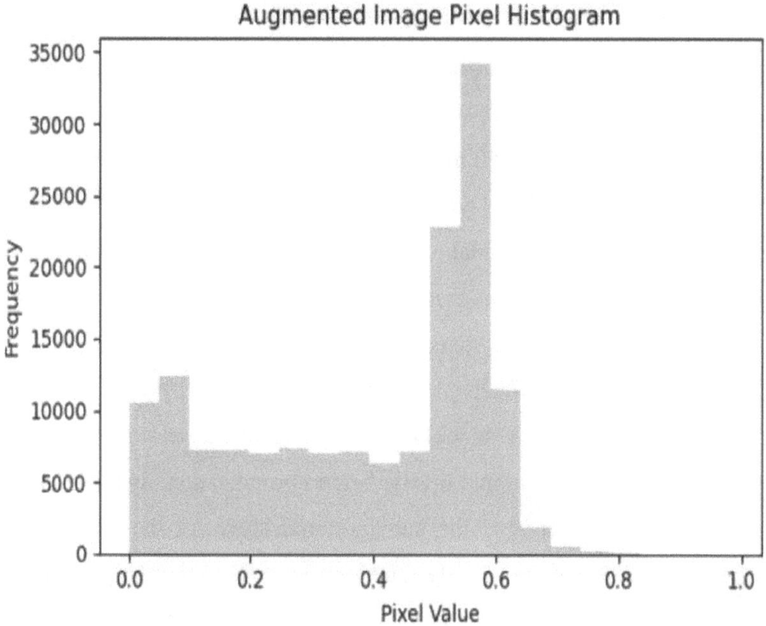

Data augmentation using shearing by range of 0.2.

Plant diseases detection A powerful deep learning model of plant disease classification is achieved through the Raspberry Pi controller relying on the features of data which is extracted by image data augmentation. We use the ResNet50 deep learning architecture on Raspberry PI to identify plant leaves and diagnose the type of disease on the individual leaf hence offers useful diagnostic measures used in agricultural crop management. The ResNet50 model is deemed powerful because of the following factors: (1) it can learn more complex features based on the data, (2) its deep framework of residual learning, which enables it to be trained on more specialized tasks and be adapted to a wide range of image classification tasks, (3) the fact that it currently uses pre-trained weights of the ImageNet dataset, which contains rich features representations obtained with the help of a huge and diverse data set, thus making it an ideal starting point in the plant disease detection process, and (4) its capacity to be trained on more specific tasks and transferred to The ResNet50 model structure proposed. Fig. 4 depicts that the ResNet50 model consists of 50 layers with a series of convolutional, pooling and fully connected layers. The residual blocks consist of a series of convolutional layers, and have shortcut connections that allow the deep neural network to learn residual functions, and alleviate the vanishing -gradient problem. The output of a residual block can be mathematically represented in equation. (1) [35, 36]. in which x is the input of the residual block, F(x, W) is the operations carried in the block, which are usually convolutional layers, batch normalization, and activation functions, and W is the weights of the convolutional layers of the block. The function is then used to define the identity functions, so that the network can learn the identity functions, by adding the input factor of the output of the function as it is called as the residual nature of the block, to result in deeper feature representations which are not degraded by any means. \[\:\vect{y}= \vect{F}\left(\vect{x},\vect{W}\right)+\:\vect{x} \] (1)**Fig. 4**

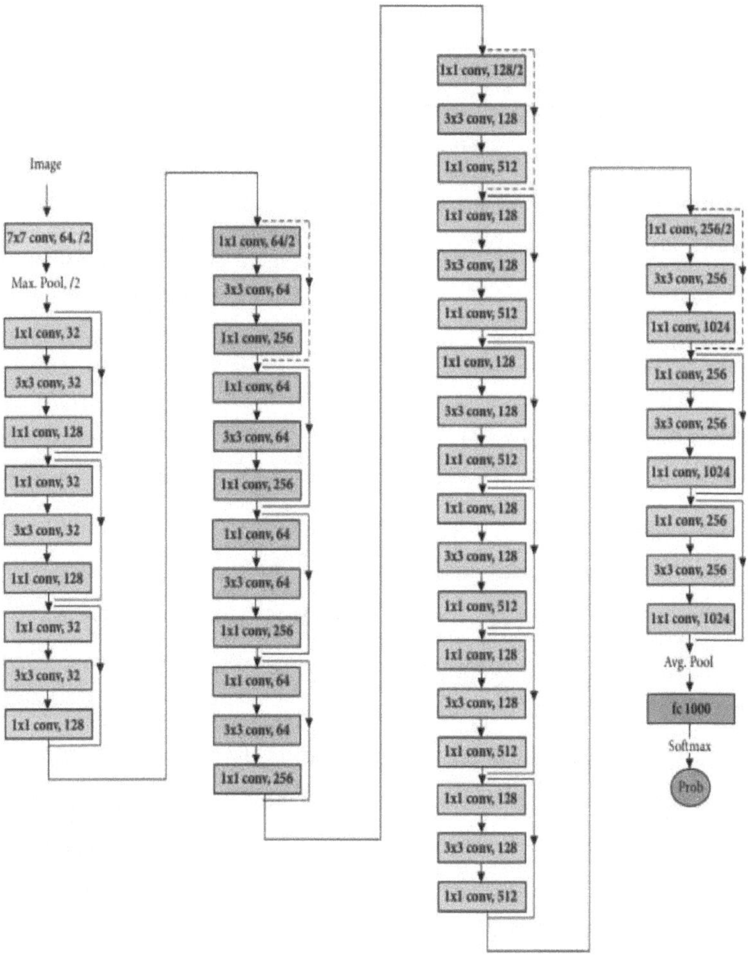

The proposed ResNet50 deep learning model structure [35].

This process downsizes the input feature maps in space, thus summarising the main statistical properties of each channel. The mathematical equation of the GAP layer is as follows in the form of equation (2) [37]: $$\:GAP\left(x\right)\:=\:\:\frac{1}{H\times\:W}\sum\:_{i=1}^{H}\sum\:_{j=1}^{W}{x}_{ij}$$ (2) H and W are the height and width of the feature map, respectively, and \: x ij is the value in the (i, j) location in the feature map. After the GAP operation, there is a fully connected layer that uses the summarised features to produce the final prediction that is subsequently subjected to a softmax layer to assist in the selection of several output classes. This layer also acts as a classifier and a detector and assigns every of the reduced feature representations of the plant image a particular class label. The fully connected layer output is presented as in the equation. (3) [38]. $$\:z\:={\text{W}}_{fc}\:y\:{b}_{fc}$$ (3) Where is the result of the GAP layer, which is the input of the fully connected layer, Wfc is a weight matrix of the fully connected layer, and bfc is the biases of the fully connected layer. The Raspberry Pi controller on the other hand would be the traditional computational unit in the heart of the irrigation system. The interface between the Raspberry Pi, the sprinklers, and the water pump has been demonstrated in figure 5 and it shows how the system water sprinklers will irrigate the plants based on the schedule and the needs of the crops.

According to Figure 5, the Raspberry Pi controller triggers the water pump to pull fresh water in the reservoir, push it in the designated pipeline and activate the sprinklers using electronic switches to supply the required amount of water during the designated irrigation schedules needed to make the plants grow. When such conditions are not met, it will switch off the water pump and the sprinklers.

Fig. 5

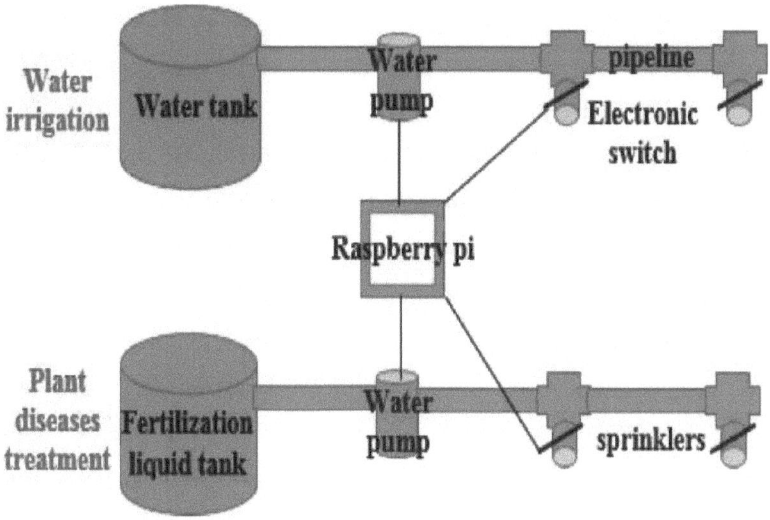

Plant diseases treatment using Raspberry pi controller.

Mobile application

Figure 6a shows the front-end of the main PLANTUM mobile application that has been implemented in the application layer of the proposed AI-IoT smart agriculture pivot. The PLANTUM mobile application allows the farmers to identify whether a leaf is healthy or diseased with a specific focus on the pepper, potato, and tomato crops. It also determines the kind of plant disease through scan operation or an upload operation. During the scan operation, the App is able to obtain any real image of a leaf and classify it into one of the 11 classes, that is, one of three healthy classes or eight disease classes as shown in Figures 6b and 6c, respectively. The image shown in figure 6b is of a healthy bell pepper that was scanned using the PLANTUM App. The App first captures the image, which is sent to the Raspberry Pi controller in the sensor field layer to classify the disease based on

ResNet 50 model. The ResNet 50 model recognizes the image taken as a healthy bell pepper. The result of this classification is further sent to the PLANTUM App and passed down to the farmer. Figure 6c presents a diseased bell pepper (bacterial spot) image which was scanned with the PLANTUM App.

Consequently, the bacterial spot on the leaf is identified by ResNet -50 as one of the eight diseased classes and forwarded to the PLANTUM App. The upload process is a duplicate of the scan process. The PLANTUM App has the capability to upload any plant images on the agricultural websites to classify them during the upload operation. Once the proposed hardware pilot was built in our experimental lab model, data augmentation, plant disease identification, plant disease curing and the mobile App were added to achieve the work of the proposed hardware pilot in the following way: • Tracking the images of leaf plants in the farm field by use of a camera attached to Raspberry Pi. • Rescaling, shearing, zooming, flipping and resizing of images to isolate accurate features using Python code on the Raspberry controller. • Using a ResNet -50 model, which was adapted, configured, and programmed with Python code on the Raspberry controller, to classify images (plant disease detection) into one of the 11 leaf-plant classes. • The possibility to control the central pivot of treating the plant disease using the Raspberry Pi controller, and actuators (sprinklers, pipelines, water tank, electronic switches).

Fig. 6

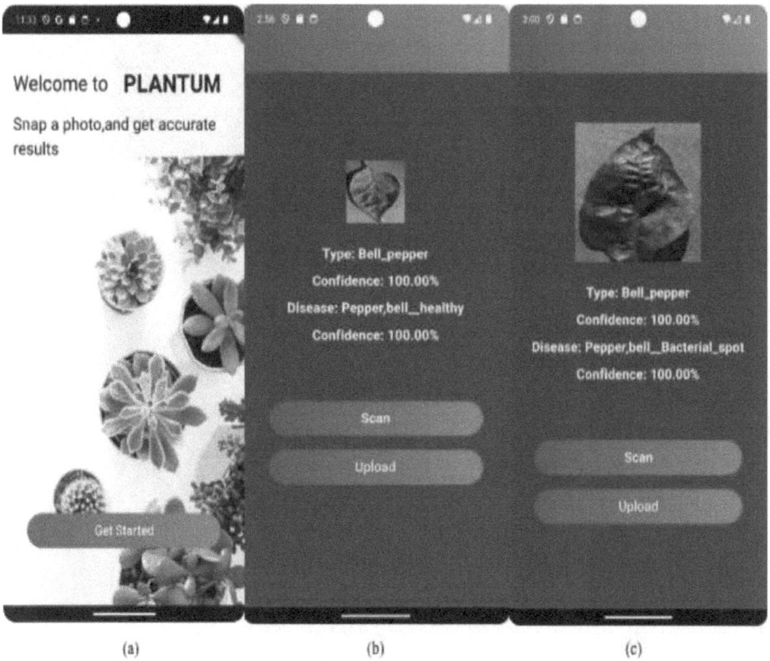

The PLANTUM App for plant diseases detection: (**a**) Main page, (**b**) scanning healthy image, (**c**) Scanning diseases image.

Figure 7 and 8 show the results of data augmentation, as is carried out by zooming and horizontal mirroring, respectively. As shown in Figure 7, the zoom transformation that has been used on the original image improves the disciplining capabilities of the model in identifying plant diseases. Figure 8 depicts how the original image can be transformed to a horizontally mirrored image. This change helps in diversifying the training set and hence enhances robustness of the model.

Fig. 7

Data augmentation using zooming.

Fig. 8

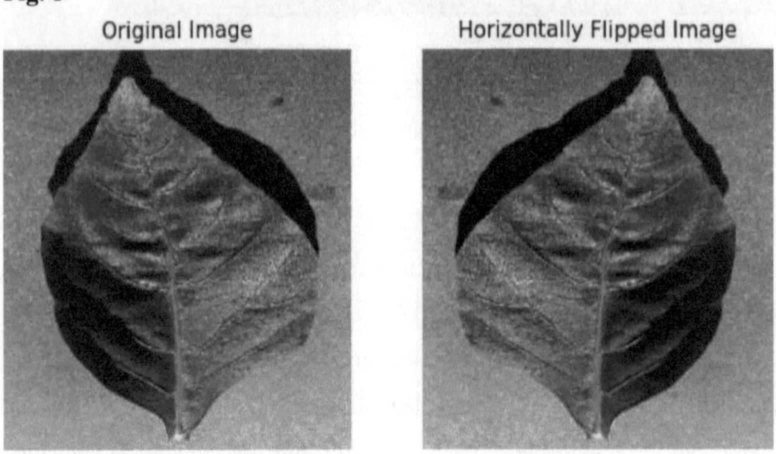

Data augmentation using horizontal flipping.

As classified report shows, the high accuracy, as well as balanced precisions, recalls and the F1-scores among the classes, support the success in classifying the plant diseases by using our model. This performance is vital in the real-world, where proper disease diagnosis is essential in proper management of crops and also yield maximisation. The classification report includes a number of assessment measures which are derived according to Equations (4) -(7) [39]. Accuracy brings in the general accuracy of the model by determining the proportion of the rightful predictions of the observations to the number of observations. It gives a rough estimate of model performance, but is not necessarily the most suitable measure in case of unbalanced data. Precision deals with predicting positive cases successfully by calculating the rate of the accurate positive predictions in respect to the number of positive cases predicted. It is especially relevant in scenarios where false positives should be kept down to a minimum e.g. in medical diagnosis or fraud detection. Recall analyzes the model of identifying correctly actual positive cases. Lastly, F1-score is a balanced score by computing the harmonic mean of precision and recall. It is particularly applicable in dealing with unequal data sets where high accuracy in itself might not tell the true performance of the model. F1 -score guarantees that both recall and precision are taken into consideration, which makes it a more comprehensive assessment measure.

Fig. 9

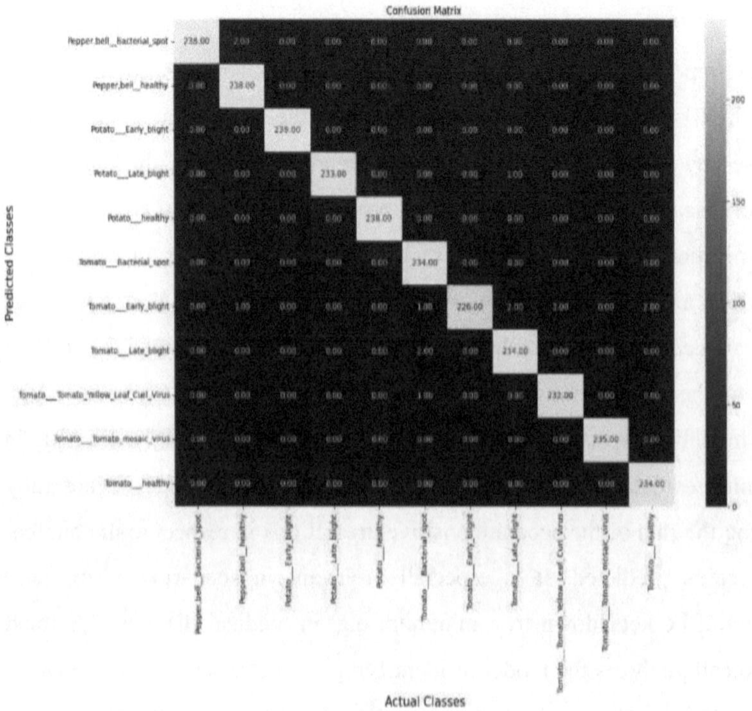

Confusion matrix of the ResNet50 model for plant diseases detection

Figure 10 shows the precision of training and testing of training ResNet50 model in 30 epochs. The orange training curve takes off at 52.34% and reaches 100% percent. Similarly, the green testing curve starts at 62.97 percent and goes up till 99.8 percent. The loss during testing and training of the ResNet50 is presented in Figure 11. The training loss reduces to 0.00001 in the 30 epochs as compared to 1.1838 and the test loss to 0.0244 as compared to 0.8635.

Fig. 10

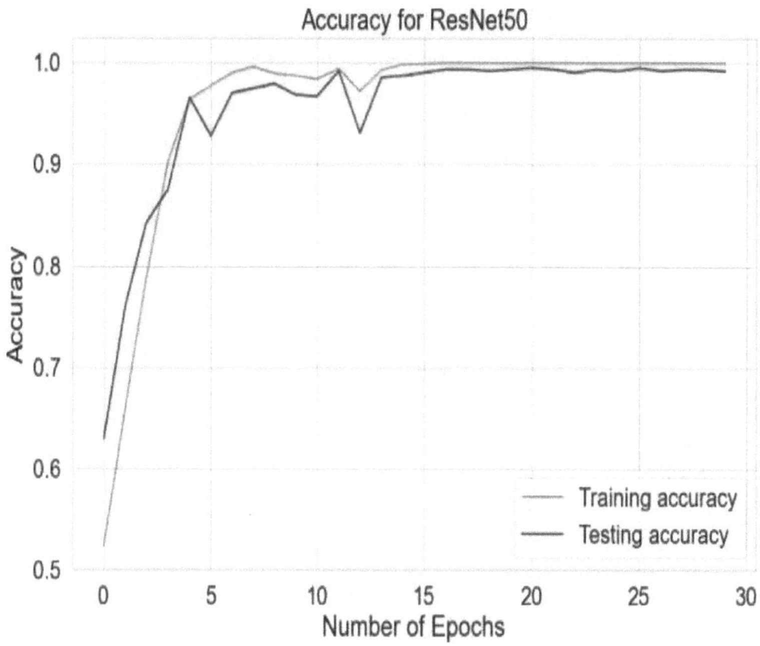

Accuracy for ResNet50.

Fig. 11

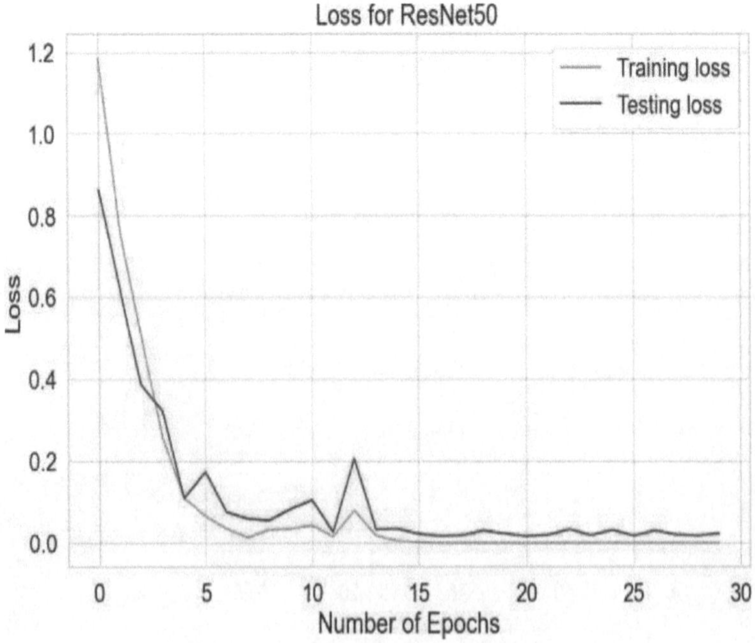

Loss for ResNet50.

Applications of AI and IoT

The world agricultural sector has been facing drastic shifts in the sphere of agriculture in the recent times that have transformed the daily routine of farm workers. With the decreasing profitability of some of the farming businesses, the augmented harvesting operations have increased, leading to the diversification of the activities in the farms. Nevertheless, the augmented industrial control has amplified mechanization and automation in the industry, thus causing a qualitative change of daily work routines of the farmers and the global farming fraternity. This has resulted in the development of the farming industry due to the emergence of new farming machinery, technology, genetically modified seeds, new and also fertilizers, and organic farming and sustainable farming methods. New technologies are also increasingly needed in the agriculture sector as a tool of sustainable development. The automation of agricultural practices is a good intervention, which has not only led to quality of the products but also minimized production costs, human labour, as well as increased environmental sustainability. Farm machinery, irrigation and fertilization, climatic control, pests and diseases, and soil fertility are also automated to enhance agricultural output [40-45]. Artificial intelligence (AI) will help farmers to yield better crops, less biomass waste, and make better decisions. AI is used to process information gathered by sensors, drones, and satellites in order to create accurate maps of the crops and soil. This information is then applied in the optimization of the use of fertilisers, irrigation and pest control.

Crop monitoring is also being used with AI technology. It analyses satellite or drone captured pictures of crops to identify problem areas. As an example, it would be able to recognize the symptoms of disease or nutritional deficiency even before they can be detected with naked eyes to enable the farmer to take corrective action before the problem gets out of hand. AI is also used to process historical data on the weather patterns, soil conditions, and crop yields in order to forecast the future. This data can be utilised to optimise the planting time, predict crop and also to establish the areas where crops are likely to fail. Most activities on the farms can be done by drones and robots that are fitted with AI features, such as planting seeds and picking crops. As a result, there is a reduction in the amount of manual labour, and the productivity of crops increases. It is possible to use AI technology to monitor the health and welfare of livestock, including the analysis of sensor data obtained on animals to identify signs of a particular disease [46-54]. The Internet of Things (IoT) is a network of devices, which detect real-time conditions and subsequently provide reactive actions; though, a significant number of IoT applications require more intricate rules to connect triggers and reactions. In the farm industry, IoT can be applied through sensors and other interconnected devices to gather real-time information and automate farming processes, resulting in higher productivity, lower expenses and higher outputs. The industry is experiencing a revolution, where AI and IoT technologies would provide farmers with real-time data and insights to enable them to make accurate decisions. Implementing AI and IoT in agriculture enhances the productivity of the farm, minimises wastage, and maximises the production of crops. IoT technology is used all over the farm to obtain real-time data on crop health, soil moisture and weather conditions. The sensors are placed on the farm and detect this data that is sent to a central database that is analysed using AI algorithms allowing the farmers to make knowledgeable decisions about irrigation, fertilisation, and other agricultural activities. AI and IoT in agriculture are able to offset the environmental effects of farming and at the same time raise crop production and reduce wastage. Through optimisation of their farm activities, farmers will be in a position to minimise pesticides and fertilisers, use less water, and reduce greenhouse emissions [55-61]. In the recent years, the AI and IoT paradigms have been widely adopted by many businesses. Precision agriculture techniques have become mostly mechanised today, with advanced AI and IoT technology. Though both the terms IoT and AI are not new, they have become extremely popular recently, mainly because of hardware improvement in the last ten years. The use of AI and IoT technology in the agricultural sector is transforming the production, saving of costs, and sustainability within the sector in a very quick manner. Additionally, the current state of precision agriculture studies has shifted towards expanding the adoption of machine learning (ML) methods to come up with creative solutions to agricultural issues. Within the framework of AI and IoT, robots, drones, remote sensors, computer and satellite imagery, constantly developing ML models, and analytical devices are used to scan crops and livestock, storage facilities, macro- and microclimate, surveying, and field mapping in real-time, offering information to farmers to make logical strategies of farm management that will not only save time and resources but also increase crop production. Such technologies are thus becoming more and more implemented in the agricultural sector to assist in easier to the environment on-farm operations as well as enhance ecological sustainability without compromising financial viability of farming businesses. [62-66].

The production of date palm (Phoenix dactylifera L.) is a significant economic practice in many arid areas, Saudi Arabia being one of them. Although date palm crop has a great economic role, it faces numerous challenges, including water shortages, pests and diseases, climate change, and lack of technology in postharvest processing. These issues require a new and progressive solution that will transform the management of crops and increase the water utilization efficiency in the production of date palms. New technologies in artificial intelligence (AI) and the Internet of Things (IoT) can revolutionize agriculture by providing live information on the quality of fruits and vegetables, tissue cultures, and crop health, soil moisture level, and weather. Furthermore, AI and IoT technologies applied in the field of date palm cultivation can bring some positive effects to the environment, such as the optimization of water, the decrease in the use of pesticides, and the quality of soil. Since the cultivation of date palms is water-intensive and may result in the overload of the local water resources, AI and IoT technologies may offer significant benefits to the environment. Among them is the optimization of the water usage, which is presented in research [6769]. With the help of soil and tree sensors, farmers will be able to check the moisture level in the soil and regulate the irrigation system. These sensors can collect data, which is then analyzed by AI algorithms and then suggested the best time and amount of water that should be applied. This will reduce the amount of water being wasted and raise the level of efficiency in the use of water especially in areas where water is scarce. The other possible benefit is that it will reduce the use of pesticides and fungicides. The IoT devices like the use of drones with cameras and sensors can also be used to monitor pest and disease attacks at an earlier stage so that farmers can take action before it spreads. Artificial intelligence can also process information on such devices to suggest the most efficient and least damaging pesticides and fungicides, which will reduce the number of harmful substances used to control and treat date palms and impact the local biodiversity favorably. The use of AI and IoT in cultivation of date palms also enhances the health of soil. With the help of the monitoring of soil bionomics, content of moisture, levels of nutrients and other factors of interest, farmers can rationalize the use of fertilizers to make sure that trees are fed with the required nutrition and that they are not over-fertilized. The strategy reduces soil erosion and improves the soundness of the soil. However, a lack of review research and book chapters dedicated to the use of AI and IoT in date palm farming, especially in Saudi Arabia, is present. The currently available literature on the use of AI and IoT in agriculture is largely focused on other crops, and little focus has been given to the date palm farming. [70-73].

Besides, majority of the previous studies have been done in the regions which are temperate and the problems facing farmers are not the same as those faced in arid areas like Saudi Arabia. Thus, the presented chapter aims to fill these knowledge gaps by discussing how the artificial intelligence and Internet of Things-based technologies, such as sensors, machine-learning algorithms, and data analytics, can be applied and benefit or pose challenges to enhancing sustainability and developing date palm farming in Saudi Arabia. The chapter also explains the challenges involved that need to be addressed to achieve the potential benefits of using AI and IoT in the production of date palm. This chapter is a great source of information to researchers and policymakers by giving them an understanding of modern trends and advances in AI and IoT in the growth of date palm in arid areas. Significance and issues of date palm farming. Date palm is a popular crop in desolate and semi-desert parts of the world, with an area under its culture taking 1.31 million hectares and producing 9.82 million tons of date fruit every year. The date palm is one of the key fruits grown in the arid areas of North Africa, Middle East, and some parts of Asia in Saudi Arabia, which provides food, nutritional and construction materials to the people and other byproducts. There are over 120 million date palm trees around the world, and over 84 million of these trees are cultivated in the Arab region including Egypt, Iraq, Saudi Arabia, Algeria, Morocco, and Tunisia among other Arab nations. The Arab nations have 70 percent of all date palm trees in the world and produce 67 percent of the world output. Furthermore, over 23 million date palm trees are planted over 152,734 hectares of land in Saudi Arabia which produces 1.57 million tons of date fruits annually. The production of date palm in Saudi Arabia is of great importance due to a number of reasons among them being its economic importance. The date palm business is a labour intensive business that favours both the male and the female labour by earning money and creating jobs. The improved job opportunities in the rural settings minimize the mass movement to urban areas. The women have a significant role especially in the propagation stages whether through in vitro or traditional means and in the postharvest stages including packaging and marketing. Date production and trade contribute to the local economies and is considered a source of income to farmers and exporters. The agricultural sector particularly date palm sector in Saudi Arabia has witnessed a high growth and sustenance over the coming decades. The export of dates is extremely dependent on many countries in the Middle East and North Africa such as Saudi Arabia, Iran, Iraq, Egypt, and Tunisia. Saudi Arabia became the largest exporter of fresh or dried dates in the world in 2021, with an export rate of about 322.84 million USD. Nonetheless, the profitability and output of the Saudi Arabia date palm growers is limited by a series of challenges such as water scarcity caused by groundwater depletion, soil degradation caused by salinization, soil erosion and desertification, loss of vegetation cover caused by overgrazing and cutting of wood to produce fuel, insect pests, diseases, inadequate processing and packaging plants and technologies, environmental contamination, and dwindling consumer demand of date fruits. [83-86].

Besides, the growers of date palms do not have proper irrigation methods such as drip irrigation and underground irrigation. Most still use traditional flood irrigation and this has led to a huge loss of water through evaporation and water run-offs. The practice also intensifies the salinization of soils thus worsening the shortage of irrigable water. In a bid to curb this problem, other farmers turn to desalinated sea water but this solution is still too costly and is beyond the power of many farmers. As a result, the government of Saudi Arabia has been channelling a substantial amount of investments in the irrigation infrastructure and new technologies to encourage judicious use of water in agriculture [87-89]. Other threats to date palm production sustainability are desertification and soil erosion. When wind or water removes the topsoil leaving the land unsuitable in growing plants, it is called soil erosion. On the other hand, desertification refers to the degradation of the previously productive land to desert-like conditions as a result of either natural or human-made activities like overgrazing, deforestation, and unsustainable agricultural activities. In Saudi Arabia, the causes of the deterioration of the soil health and the development of desertification are the changes in the climate, overuse of groundwater, and unproductive farming techniques. It is especially susceptible to these challenges because of the dry climate and a shortage of water in the country [90-96]. The pest and disease-related issues are another problem that date palm farmers in Saudi Arabia are facing. Various pests such as red palm weevils, longhorn beetles, dust mites, fruit flies, scale insects and fungal pathogens can be found in the date palms. An example of an invasive species that feeds on young and mature palm trees is the red palm weevil that destroys the vascular system of trees. Larvae tunnel into the trunk and feed on soft tissues causing tunnels to undermine the integrity of the structure. The leaves with the virus will wilt and turn yellow and in most cases, the trees will die [97-105]. In Saudi Arabia, a major industry is date palm farming where postharvest losses are approximated to be about 10% of the total. One of the main reasons of such losses is the lack of proper postharvest facilities. It is necessary to store, process, and pack the fruits properly to avoid spoilage, losses, and fruit quality. But high cost, lack of expertise and inadequate infrastructures among other factors have hampered the use of thereof postharvest facilities by many of the date palm growers.

[106- 110].

Skill manpower shortage is also another problem that afflicts Saudi Arabian date palm growers. Numerous date palm farmers rely on migrant workers during harvest time and other cultural activities like pollination, pruning, fertilisation, irrigation, application of chemicals and so on. Nonetheless, changed immigration policy and geopolitical tensions have made it hard to get skilled labour where needed by farmers. This results in labour problems at peak seasons that cause low yields and losses to farmers. Therefore, labour shortages in case of skilled labour during moments of critical situations reduce production and cause financial losses to farmers. AI and IoT technologies Artificial Intelligence is a set of technologies, which allow computers to complete a significant number of diverse and complicated tasks, such as image recognition, verbal and written language understanding and translation, data processing, providing recommendations, etc. It has transformed agriculture to make it efficient, sustainable and profitable. The Internet of Things is also being embraced by many businesses, offering easy methods of collecting and processing technical information to detect and enhance the effectiveness of different daily processes. The technological revolution introduces new challenges and problems regarding the present-day IoT technologies. New technologies like AI, 5G and blockchain present solutions to these problems. With the combination of IoT and AI, we will be able to create intelligent machines capable of simplifying repetitive work and also making autonomous and intelligent decisions. IoT and AI are two technologies that have made an impact on the computing world as two of the most important technological advancements, which have radically altered the way we interact with machines and the environment. It is estimated that by 2025 some 64 billion AI and IoT devices will be on offer. The AI and IoT are compatible and are on the edge of the newest innovations that are influencing information technology industry. The redesign of traditional solutions, which is brought by this partnership, has helped the industrial and agribusiness industries. Machine learning (ML) is a branch of AI that entails the use of training algorithms to predict or make choices using training data. ML is a fast growing discipline that has many applications and chances of scientific innovation. It is an aspect of AI that designs algorithms and statistical models able to help computer systems automatically increase their performance in a particular task as they gain experience. ML aims at developing smart systems that have the ability to learn through data and make predictions or judgments without a specific programming. ML as a technique used in this context is an application of analyzing sensor data and predicting future events in the context of IoT. The use of deep learning as a powerful ML approach has become a thing of the last few years. Deep learning is a technique that uses multi-layer neural networks that are trained to obtain hierarchical representations of data. Thus, there has been significant advancement in areas like speech recognition, natural language processing and image recognition. ML has its applications in different industries such as agriculture, healthcare, finance, marketing and transportation among others. ML has been used in farming to weed, detect pests and diseases, predict crop yields and quality, collect data, generate insights and predict livestock production. In healthcare, it is able to identify new medicines and diagnoses. It may be used to evaluate risk and fraud in finance. It may be used in marketing to allow personalized advertisement and customer segmentation. It can be used in transportation to aid autonomous driving and predicting the traffic. [114- 120].

Natural language processing (NLP) is a branch of artificial intelligence that is used in the interaction between computers and humans with the help of natural language. NLP cannot be ignored in the current AI systems since it gives machines the opportunity to comprehend, interpret, and produce human language. The development of smart homes, smart cities, and other intelligent systems can now enjoy the application of NLP to the Internet-of-Things (IoT) devices. The integration of voice -recognition and natural-language-understanding features enables the user to easily and efficiently interact with the IoT devices. The use of

NLP in the field of agriculture is an example of weather-data analysis. By assessing the weather patterns and forecasts, farmers can be in a position to know the best time to plant and harvest crops, how to irrigate their farm, and the method to control pests and diseases. Through the help of NLP algorithms, farmers can observe the crop-growth and development rates, and the level of soil-moisture. NLP models can also be used to forecast demand and assess market trends to assist the farmers in choosing the most appropriate crops to be grown and also in deciding the best time when to sell their crops. NLP algorithms help farmers to locate new customers and communicate with them to achieve higher prices. NLP can be applied in the area of consumer-farmer communication. By examining the feedback of the customers, farmers are in a position to identify the preferences of the consumers. Through NLP algorithms, the farmers will be able to communicate better with each other, exchange best farming tips and together with scientists in conducting research projects [121-125]. Computer vision is a component of artificial intelligence that allows computing systems to extract valuable information about visual data, digital images, and videos, and make decisions or offer suggestions. This technology is commonly used in IoT to enable the devices to sense and discern physical objects by recognizing objects, faces and gestures and thus making them more intelligent and reactive. By exploiting deep-learning algorithms, computer-vision systems are able to detect objects in real-time, and label them based on some of the given classes. The capability has been used in other cases of the IoT such as autonomous vehicles, security cameras, and smart-home devices. Computer-vision systems are also able to produce useful data about the environment through images taken by the sensors or cameras. To give an illustration, they have the capability of identifying anomalies or a change of a scene, observe or monitor traffic flow, or even the movement of people or objects.

Gesture recognition and human-computer interaction are other technologies that IoT has used. With the help of computer-vision, users can engage in interactions with devices in a more natural and intuitive way through the analysis of hand movements and gestures. Computer-vision technology is being used to automate and optimise many of the agricultural processes. Some of the applications of computer-vision technology in the agricultural industry are crop tracking, estimate crop yields, and detection of insects and diseases [126-130]. Robotics technology is one of the branches of engineering that is direct to the development of AI and the IoT by means of designing and programming of robots. It is also used in IoT to develop intelligent machines that can communicate as well as be interconnected. The use of machines in robotics is based on the fact that they are either in dangerous situations or too challenging to be done by humans. The way of how we communicate with machines has been changed dramatically with the implementation of robotics technology into AI and IoT, and this has made them efficient, reliable, and autonomous. These robots work on 24/7 basis, which means they perform data-entry and processing which tends to minimize errors and increase efficiency. Robotization of processes has become common in various industries, such as agriculture, manufacturing, health and wellbeing, logistics, and finance. Another application of robotics technology in the field of AI is the development of autonomous robots. Intelligent robots which have sensors, cameras and other advanced technologies can be used to move around the world even without the presence of human beings. The use of robotics technology in agriculture is rapidly being utilized in order to enhance productivity and efficiency of farming activities. The use of robots in the field of agriculture is said to bring numerous advantages, including better crop, reduced labour expenses, and increased precision [13,134]. Edge computing is an example of decentralised computing that brings computation and storage of data nearer to the location of data generation and this increases response time and reduces the amount of bandwidth used. Edge computing is another concept in the IoT where sensor information gets processed in real time at the network edge. The AI models based on machine learning are usually developed with high amounts of data. This training may be performed locally through edge computing and hence less volume of data is required to be transmitted to the centralized servers. Storing sensitive data locally instead of sending it through the internet helps reduce network congestion, improves privacy and security, and improves reliability as well, as edge computing helps reduce network congestion. Real time running of AI models can also be implemented based on edge computing which can result in quick response and decision making. An example is that edge computing can process sensor data on a localised region on real time and detect potential issues before they were overpowered.[135-140].

Furthermore, developers can increase or decrease their machine learning models according to the demand through cloud-based services. Besides, cloud computing presents a safe place to process and store confidential information created by the Internet of Things and artificial intelligence apps. Cloud service providers have taken many security risks such as firewalls, access controls, and encryption to ensure that their customers information is not accessible to attacks by cyber criminals. One of the previous researches highlighted the benefits of using the Internet of Things in the agricultural sector, especially irrigation management. The automated system of scheduling irrigation created by Mohammed et al. based on a cloud-based IoT platform impacted the yield of date palms and the efficiency of the water utilization through an If This Then That interface and the Ubidots platform in a positive way. The main elements of the cloud-based IoT platform used in monitoring meteorological variables in real-time and controlling the irrigation schedule can be seen in Figure 12. The diagram illustrates how to send data sensors to the cloud platform and how to control the irrigation system using Arduino UNO board and an IFTTT interface. Sensors also gathered meteorological variables and conveyed them in real time to the microcontroller Arduino through a Wi-Fi module. The microcontroller subsequently sent the information to the Ubidots platform, where the users were able to see the real-time information via a graphical user interface via a private channel. Contactor electronic relays, electronic valves and irrigation pumps were also controlled using the IFTTT interface to schedule irrigation water in the system.

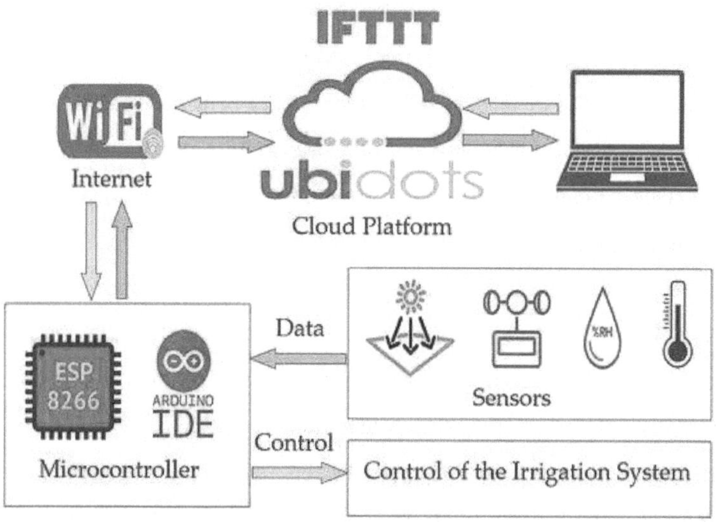

Figure 12.

A simple diagram for the components of the cloud-based IoT system employed for real-time monitoring of the meteorological parameters of the study area and managing water scheduling for date palms.

In the article by Mohammed et al. [59], cloud-based IoT-based solutions were used to manage a modern subsurface irrigation system in date palm farms in the dry area of Saudi Arabia, enhancing the management of irrigation. They made and built a fully autonomous Controlled Subsurface Irrigation System (CSIS) and verified its functions to monitor the irrigation process remotely. A flow water management system that involved the use of a sophisticated cloud computing and internet of things was applied to control water to date palms. The notification to the user can be implemented automatically through SMS or email message. Subsurface irrigation system in date palm field can have the optimum volume of water per tree. In this research, the researcher will employ a methodology of designing and deploying a cloud-based CSIS to date palm trees. CSIS is an IoT-based service, which involves use of cloud computing and other sensors to monitor and manage date palm subsurface irrigation. Subsurface irrigation scheduling (S-BIS) that was sensor based was adopted. The quantity of water is scheduleable basing on the information obtained through the sensors. The sensor data is measured and uploaded to the ThingSpeak cloud platform to be analyzed and the resultant decision is sent to the subsurface irrigation system. Figure 13 depicts the proposed system which involves direct volumetric water content measurement on which irrigation decisions are made and other parameters measured are ambient air temperature, relative humidity, sunlight, wind velocity and the rate of water flow per minute. The findings showed that automatic irrigation of date palm trees under S-BIS was more efficient compared to automatic irrigation based on time (T-BIS). S-BIS and T-BIS experienced a 64.1 per cent and 61.2 per cent reduction in the amount of irrigation water used per alternative compared with the use of traditional surface irrigation (TSI). The total irrigation water per CSIS with the S -BIS process was 21.04, 22.76, and 58.71 m 3 per tree in a year, respectively. Combined with the S-BIS methodology, the developed CSIS strategy provides positive results related to irrigation water management and further enhancement of the date palm fruit production on the arid areas [59].

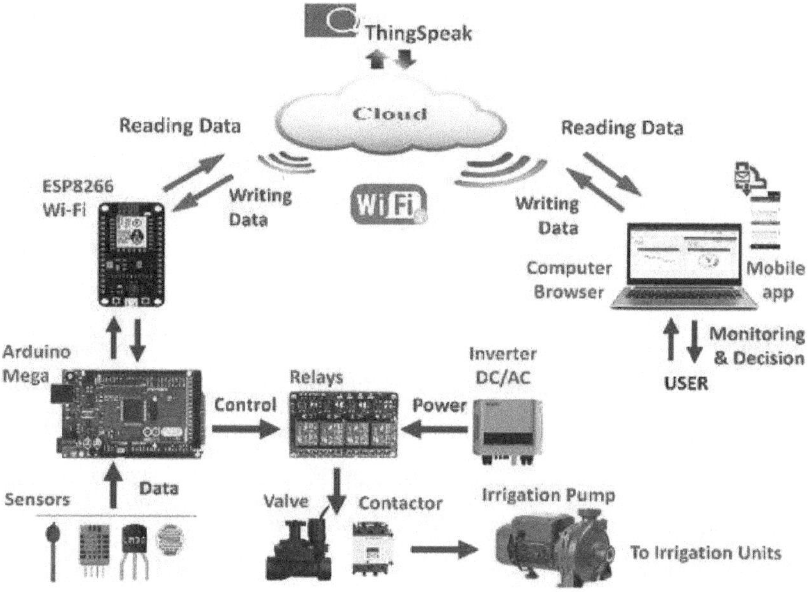

Figure 13.

A schematic diagram for the IoT-based control system for a smart subsurface irrigation for enhancing irrigation management of date palm.

The IoT-based control system is effective in scheduling the irrigation water to be applied to date palms in different times depending on the information provided by different sensors. The system gathers the measurements of these sensors and sends the information to ThingSpeak cloud to conduct analyses, process the data on the cloud, make inferences, and transform the inferences to the subsurface irrigation system. The advantages of the IoT technology application in the current study are the more efficient water management process, better crop yield, and less wasted water. The system is also remotely monitored and controlled, and thus manual labor is less required and the precision of the date palm irrigation control is elevated. The SEVAS is an automated system that was designed by Mohammed et al. to acclimatize plantlets in a tissue culture by using an IoT system. The proposed solution uses the IoT technology to monitor and regulate the environmental conditions of the glasshouse such as temperature, humidity, and light intensity. The feedback mechanism is also implemented in the system where the environmental conditions are as well adjusted depending on the real-time data obtained by the sensors. The benefits of using the IoT technology in this study include the real-time monitoring of crops and microclimate, field surveying and mapping, and supplying the farmers with information to use sound measures in the farm management. The strategies are geared towards increased efficiency, time and resources saving in addition to high crop yield. Moreover, the IoT is increasingly becoming a trend in agriculture and, thus, enabling the implementation of ecologically friendly on-farm practices and enhancing better ecological sustainability. Figure 3 gives a summary of the IoT-based monitoring and control system of the SEVAS in the research. The diagram demonstrates the major elements of the system and includes sensors, a microcontroller, an internet connection, the IoT platform that is based on the cloud, control equipment, and web-based applications. The components are interrelated and synchronized in order to provide control and monitoring capabilities. The figure also illustrates the path of the data of the sensors to the cloud-based IoT platform and control devices that monitor the variables of microclimate in the E-VAS system. The data and control signals are then stored and analyzed in the cloud-based IoT platform and sent to the control devices which consist of relays to control the heating unit, cooling units, ultrasonic humidifier, and water pump and irrigation valves. The control devices vary the SEVAS microclimate factors according to the data of the sensors and the control signals through the cloud-based IoT platform.[79].

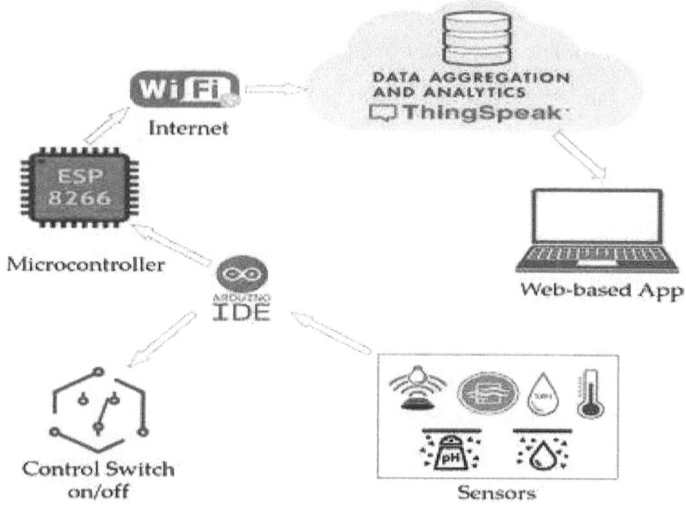

Figure 14.

A simple diagram for the components of the IoT-based controlling and monitoring system of the SEVAS for tissue culture plantlets.

Conclusion

The chapter proposes the detection and treatment of plant diseases based on a suggested AI-IoT Smart Agriculture Pivot that was conceived to tackle the issue of crop production and food shortage in the agricultural sector. Furthermore, the paper presents several improvement methods and technologies, as well as associated previous researches, which are widely used to detect plant diseases. Moreover, the paper also singles out some of the most widespread agricultural machines, including UAVs, robotics, and pivots, as the remedies against plant maladies. The development of the proposed AI-IoT Smart Agriculture Pivot follows the structure of Python code, implemented, and programmed in the Raspberry controller to detect and treat plant diseases. The proposal is going to be programmed with data augmentation and ResNet50 deep-learning model to train, validate and test the augmented images obtained in NPDD repository. One concludes that the ResNet50 model achieves a high accuracy in classification of eleven classes of plants, which is higher than the other deep-learning models but also higher than other prior works in detecting plant diseases. The experimental findings showed that ResNet -50 got the testing accuracy of 99.8 and the F1-score, recall, and the precision of the ResNet50 model were 99.91, 99.92, and 100, respectively. It has come up with a mobile application that will assist farmers with professional recommendations and instructions on how to detect and treat the disease of their plants. In addition, the paper shows a system of spraying fertilizers against the identified plant disease which is automated. The implemented hardware pilot can be controlled by the RaspberryPad which can spray liquid on farmland to treat plant diseases. In our future work, we will improve the process of water irrigation in the proposed AI-IoT Smart Agriculture Pivot compared to the traditional central pivot to save the water resources. The suggested pivot will also create pest-attack control systems and their respective treatment.

The introduction of artificial intelligence (AI) and the Internet of Things (IoT) in the growth of date palm in Saudi Arabia proves that it has a considerable potential to enhance the productivity, sustainability, and quality control of its cultivation. There has been a rise in the demand to implement AI and IoT technologies to grow and manage date palm trees in the Kingdom. Several researches have examined how AI and IoT will be applied in different aspects of date palm production, such as pest management, postharvest quality control, yield improvement, and the spatial mapping of date palm orchards. The application of AI and IoT provides the advantages of higher crop production, better pest management, and higher quality of harvests. These technologies help to analyze data in real time and monitor it, which in turn allows farmers to make sound choices and distribute the resources they have more effectively. The combination of AI and IoT can help make the process of date palm cultivation more sustainable by minimizing the use of water and eliminating the effects of climate change. However, it is imperative to note that, although the AI and IoT technologies have potential, there are still a number of issues that have to be overcome. These issues are the need to collect and analyze data in a robust way, establish the data security and privacy, and eliminate the digital divide in rural areas. The potential of date palm cultivation using AI and IoT still needs to be fully utilized through further research and development. Continued research and development regarding the same can help to improve the date palm farming and benefit the entire agricultural industry.

References

1. Fróna, D., Szenderák, J. & Rákos, M. H. The challenge of feeding the world. *Sustainability* **11**(20), 5816 (2019).
2. Ray, D. K., Mueller, N. D., West, P. C. & Foley, J. A. Yield trends are insufficient to double global crop production by 2050. *PLoS One* **8**(6), 1–8 (2013).

3. Ristaino, J. B. et al. The persistent threat of emerging plant disease pandemics to global food security. *Natl. Library Med.* **118**(23), e2022239118 (2021).
4. Hassan, S. M., Maji, A. K., Jasiński, M., Leonowicz, Z. & Jasińska, E. Identification of plant-leaf diseases using CNN and transfer-learning approach. *Electronics* **10**(12), 1388 (2021).
5. Meena, S. D., Kumar, K. A. Y., Mandava, D., Sri, K. B., Panda, L. et al., Plant diseases detection using transfer learning. In *Proceedings of fourth international conference on computer and communication technologies*, pp. 1–16 (2023).
6. Ali, A., Yaseen, M., Aljanabi, M. & Abed, S. Transfer learning: A new promising techniques. *Mesopotamian J. Big Data* **2023**, 29–30 (2023).
7. Deore, B. D., Kasar, P., Bhamare, M. A. & Sahane, M. Drones for smart agriculture: A technical report. *Int. J. Res. Appl. Sci. Eng. Technol. (IJRASET)* **6**(4), 341–346 (2018).
8. Unpaprom, Y., Dussadeeb, N. & Ramaraj, R. Modern agriculture drones, *The Development of Smart Farmers*, pp.13–19 (2018).
9. Talaviya, T., Shah, D., Patel, N., Yagnik, H. & Shah, M. Implementation of artificial intelligence in agriculture for optimisation of irrigation and application of pesticides and herbicides. *Artif. Intell. Agric.* **4**, 58–73 (2020).
10. Qazi, S., Khawaja, B. A. & Farooq, Q. U. IoT-equipped and AI-enabled next generation smart agriculture: A critical review, current challenges and future trends. *IEEE Access* **10**, 21219–21235 (2022).
11. Xiang, H. & Tian, L. Development of a low-cost agricultural remote sensing system based on an autonomous unmanned aerial vehicle (UAV). *Biosys. Eng.* **108**(2), 174–190 (2011).

12. Boursianis, A. D. et al. Internet of Things (IoT) and agricultural unmanned aerial vehicles (UAVs) in smart farming: A comprehensive review. *Internet Things* **18**, 100187 (2020).
13. Friha, O., Ferrag, M. A., Shu, L., Maglaras, L. & Wang, X. Internet of Things for the future of smart agriculture: A comprehensive survey of emerging technologies. *IEEE/CAA J. Automatica Sinica* **8**(4), 718–752 (2021).
14. Kumar, A. et al. Artificial intelligence, internet of Things (IoT) and smart agriculture for sustainable farming: A review. *Ann. Plant Sci.* **11**(11), 5512–5564 (2022).
15. Alzubi, A. A. & Galyna, K. Artificial intelligence and internet of Things for sustainable farming and smart agriculture. *IEEE Access* **11**, 78686–78692 (2023).
16. Barnett, J., Duke, M., Au, C. K. & Lim, S. H. Work distribution of multiple cartesian robot arms for kiwifruit harvesting. *Comput. Electron. Agric.* **169**, 105202 (2020).
17. Steward, B. L., Gai, J. Y. & Tang, L. The use of agricultural robots in weed management and control. In *Robotics and automation for improving agriculture*, Cambridge, pp. 1–25 (UK: Burleigh Dodds Science Publishing, 2019)
18. Santos, P. G. D. et al. Fleets of robots for environmentally-safe pest control in agriculture. *Precision Agric.* **18**(4), 574–614 (2017).
19. Ponce, D. F. Y., Salcedo, J. V., Montalvo, P. D. R. & Sanchis, J. Mobile robotics in smart farming: Current trends and applications. *Front. Artif. Intell.* **6**, 1–13 (2023).

20. Banerjee, S. & Mondal, A. An intelligent approach towards plant leaf disease detection through different convolutional neural Networks. *Int. J. Intell. Syst. Appl. Eng.* **12**(2), 536–546 (2024).
21. Sun, X. et al. Research on plant disease identification based on CNN. *Cognitive Robot.* **2**, 155–163 (2022).
22. Krishnan, R. & Julie, E. Computer aided detection of leaf disease in agriculture using convolution neural network-based squeeze and excitation network. *Automatika* **64**(4), 1038–1053 (2023).
23. Jerome, N., Jothiraj, S., Kandasamy, S. & Ramachandran, D. An effective approach for plant disease detection using assessment-based convolutional neural networks (A-CNN). *J. Adv. Res. Appl. Sci. Eng. Technol.* **31**(3), 155–172 (2023).
24. Lee, T. Y., Lin, I. A., Yu, J. Y., Yang, J. M. & Chang, Y. C. High efficiency disease detection for potato leaf with convolutional neural network. *SN Comput. Sci.* **2**(297), 1–11 (2021).
25. Agarwal, M., Sinha, A., Gupta, S., Mishra, D. & Mishra, R. Potato crop disease classification using convolutional neural network. *Smart Syst. IoT: Innovat. Comput.* **141**, 391–400 (2019).
26. Mustafa, H. et al. Pepper bell leaf disease detection and classification using optimized convolutional neural network. *Multimedia Tools Appl.* **82**, 12065–12080 (2023).
27. Mahamud, F. et al. Bell pepper leaf disease classification using convolutional neural network. *Conf. Intell. Comput. Optimiz.* **569**, 75–86 (2022).
28. Shilpa, A., Muneeswaran, V. & Rathinam, D. D. K. A precise and autonomous irrigation system for agriculture: IoT based self-propelled center

pivot irrigation system. In *International conference on advanced computing & communication systems (ICACCS)*, pp. 533–538, Coimbatore, India, (2019).

29. Ragab, M. A., Badreldeen, M. M., Sedhom, A. & Mamdouh, W. M. IoT based smart irrigation system. *Int. J. Ind. Sustain. Dev. (IJISD)* **3**(1), 76–86 (2022).
30. Matilla, D. M., Murciego, Á. L., Bravo, D. M. J., Mendes, A. S. & Leithardt, V. R. Q. Low cost center pivot irrigation monitoring systems based on IoT and LoRaWAN technologies. In *2020 IEEE international workshop on metrology for agriculture and forestry (MetroAgriFor)*, pp. 262–267, Trento, Italy, (2020).
31. Debauche, O. et al. Edge AI-IoT pivot irrigation, plant diseases, and pests identification. *Int. Conf. Emerg. Ubiquitous Syst. Pervasive Netw. (EUSPN)* **177**, 40–48 (2020).
32. Debauche, O., Mahmoudi, S., Mahmoudi, S. A., Manneback, P. & Lebeau, F. A new edge Architecture for AI-IoT services deployment. *Int. Conf. Mobile Syst. Pervasive Comput. (MobiSPC)* **175**, 10–19 (2020).
33. https://cdn.shopify.com/s/files/1/0672/9409/files/Datasheet_SIM800L.pdf?v= 1601544539. Access: June 2024.
34. https://datasheets.raspberrypi.com/rpi4/raspberry-pi-4-datasheet.pdf. Access: June 2024.
35. Bhatti, A. et al. Explicit content detection system: An approach towards a safe and ethical environment. *Appl. Comput. Intell. Soft Comput.* **2018**, 1–13 (2018).
36. He, K., Zhang, X., Ren, S. & Sun, J. Deep residual learning for image recognition. In *IEEE Conference on computer vision and pattern recognition (CVPR)*, pp. 770–778, Las Vegas, NV, USA, (2016).

37. Krishnana, R. & Julie, E. Computer aided detection of leaf disease in agriculture using convolution neural network based squeeze and excitation network. *Automatika* **64**(4), 1038–1053 (2023).

38. He, K., Zhang, X., Ren, S. & Sun, J. Delving deep into rectifiers: surpassing human-level performance on imagenet classification. In *2015 IEEE International Conference on Computer Vision (ICCV)*, pp. 1026–1034, Santiago, Chile, (2015).

39. Mohsen, S. et al. Brain tumor classification using hybrid single image super-resolution technique with resnext101_32 × 8d and vgg19 pre-trained models. *IEEE Access* **11**, 55582–55595 (2023).

40. Edan Y, Han S, Kondo N. Automation in agriculture. In: Nof S, editor. Springer Handbook of Automation. Berlin, Heidelberg: Springer Berlin Heidelberg; 2009. pp. 1095-1128. DOI: 10.1007/978-3-540-78831-7_63

41. Elhassan Ahmed OM, Osman AA, Awadalkarim SD. A design of an automated fertigation system using IoT. In: 2018 International Conference on Computer, Control, Electrical, and Electronics Engineering (ICCCEEE). Khartoum, Sudan: IEEE; 2018. pp. 1-5. Available from: https://ieeexplore.ieee.org/document/8515772/

42. Lavanya G, Rani C, Ganeshkumar P. An automated low cost IoT based fertilizer intimation system for smart agriculture. Sustainable Computing: Informatics and Systems. 2020;28:100300. Available from: https://linkinghub.elsevier.com/retrieve/pii/S2210537918302567

43. Ben Ayed R, Hanana M. Artificial intelligence to improve the food and agriculture sector. Khan R, editor. Journal of Food Quality. 2021;22:1-7. Available from: https://www.hindawi.com/journals/jfq/2021/5584754/

44. Boursianis AD, Papadopoulou MS, Diamantoulakis P, Liopa-Tsakalidi A, Barouchas P, Salahas G, et al. Internet of things (IoT) and agricultural unmanned aerial vehicles (UAVs) in smart farming: A comprehensive review. Internet of Things. 2022;18:100187. Available from: https://linkinghub.elsevier.com/retrieve/pii/S2542660520300238
45. Sharma A, Georgi M, Tregubenko M, Tselykh A, Tselykh A. Enabling smart agriculture by implementing artificial intelligence and embedded sensing. Computers and Industrial Engineering. 2022;165:107936. Available from: https://linkinghub.elsevier.com/retrieve/pii/S0360835222000067
46. Jha K, Doshi A, Patel P, Shah M. A comprehensive review on automation in agriculture using artificial intelligence. Artificial Intelligence in Agriculture. 2019;2:1-12. Available from: https://linkinghub.elsevier.com/retrieve/pii/S2589721719300182
47. Suleimenov IE, Vitulyova YS, Bakirov AS, Gabrielyan OA. Artificial intelligence. In: Proceedings of the 2020 6th International Conference on Computer and Technology Applications. New York, NY, USA: ACM; 2020. pp. 22-25. DOI: 10.1145/3397125.3397141
48. Neethirajan S. The role of sensors, big data and machine learning in modern animal farming. Sensing and Bio-Sensing Research. 2020;29:100367. Available from: https://linkinghub.elsevier.com/retrieve/pii/S2214180420301343
49. Liu SY. Artificial intelligence (AI) in agriculture. IT Professional. 2020;22(3):14-15
50. 11.Linaza MT, Posada J, Bund J, Eisert P, Quartulli M, Döllner J, et al. Data-driven artificial intelligence applications for sustainable precision agriculture. Agronomy. 2021;11(6):1227. DOI: 2073-4395/11/6/1227

51. Basri R, Islam F, Shorif SB, Uddin MS. Robots and drones in agriculture—A survey. In: Computer Vision and Machine Learning in Agriculture. Singapore: Springer; 2021. pp. 9-29. DOI: 10.1007/978-981-33-6424-0_2

52. Bogue R. Robots poised to transform agriculture. The International Journal of Robotics Research and Application. 2021;48(5):637-642

53. Neethirajan S. Affective state recognition in livestock—Artificial intelligence approaches. Animals. 2022;12(6):759. DOI: 2076-2615/12/6/759

54. Wakchaure M, Patle BK, Mahindrakar AK. Application of AI techniques and robotics in agriculture: A review. Artificial Intelligence in the Life Sciences. 2023;3:100057. Available from: https://linkinghub.elsevier.com/retrieve/pii/S2667318523000016

55. Kim WS, Lee WS, Kim YJ. A review of the applications of the internet of things (IoT) for agricultural automation. Journal of Biosystems Engineering. 2020;45(4):385-400

56. Parvez B, Haidri RA, Kumar VJ. IoT in agriculture. In: 2020 International Conference on Computational Performance Evaluation (ComPE). Shillong, India: IEEE; 2020. pp. 844-847. Available from: https://ieeexplore.ieee.org/document/9200035/

57. Vangala A, Das AK, Kumar N, Alazab M. Smart secure sensing for IoT-based agriculture: Blockchain perspective. IEEE Sensors Journal. 2021;21(16):17591-17607

58. Sagheer A, Mohammed M, Riad K, Alhajhoj M. A cloud-based IoT platform for precision control of soilless greenhouse cultivation. Sensors (Switzerland). 2021;21(1):1-29. DOI: 1424-8220/21/1/223

59. Mohammed M, Riad K, Alqahtani N. Efficient iot-based control for a smart subsurface irrigation system to enhance irrigation management of date palm. Sensors. 2021;21(12):3942. DOI: 10.3390/s21123942
60. Kashyap PK, Kumar S, Jaiswal A, Prasad M, Gandomi AH. Towards precision agriculture: IoT-enabled intelligent irrigation systems using deep learning neural network. IEEE Sensors Journal. 2021;21(16):17479-17491. Available from: https://ieeexplore.ieee.org/document/9388691/
61. Kailasam S, Achanta SDM, Rama Koteswara Rao P, Vatambeti R, Kayam S. An IoT-based agriculture maintenance using pervasive computing with machine learning technique. International Journal of Intelligent Computing and Cybernetics. 2022;15(2):184-197
62. Zhao J-c, Zhang J-f, Yu F, Guo J-x, Zhao JC, Zhang JF, et al. The study and application of the IOT technology in agriculture. In: 2010 3rd International Conference on Computer Science and Information Technology. Chengdu, China: IEEE; 2010. pp. 462-465. Available from: http://ieeexplore.ieee.org/document/5565120/
63. Patil VC, Al-Gaadi KA, Biradar DP, Rangaswamy M. Internet of things (Iot) and cloud computing for agriculture: An overview. In: Proceedings of Agro-Informatics and Precision Agriculture (AIPA 2012), India. 1 Aug 2012. pp. 292-296
64. Muangprathub J, Boonnam N, Kajornkasirat S, Lekbangpong N, Wanichsombat A, Nillaor P. IoT and agriculture data analysis for smart farm. Computers and Electronics in Agriculture. 2019;156:467-474. Available from: https://www.sciencedirect.com/science/article/abs/pii/S0168169918308913

65. Salam A. Internet of things for sustainable community development: Introduction and overview. In: Salam A, editor. Internet of Things. Cham, Switzerland: Springer International Publishing; 2020. pp. 1-31. DOI: 10.1007/978-3-030-35291-2_1
66. Rehman A, Saba T, Kashif M, Fati SM, Bahaj SA, Chaudhry H. A revisit of internet of things technologies for monitoring and control strategies in smart agriculture. Agronomy. 2022;12(1):127. DOI: 2073-4395/12/1/127
67. Zaid A, De Wet PF. Date palm cultivation. In: FAO Plant Production and Protection Paper 156 Rev1. Rome, Italy: FAO; 2002. pp. 1-28
68. Mohammed M, Hamdoun H, Sagheer A. Toward sustainable farming: Implementing artificial intelligence to predict optimum water and energy requirements for sensor-based micro irrigation systems powered by solar PV. Agronomy. 2023;13(4):1081. DOI: 2073-4395/13/4/1081
69. Mohammed M, Riad K, Alqahtani N. Design of a smart IoT-based control system for remotely managing cold storage facilities. Sensors. 2022;22(13):4680. DOI: 1424-8220/22/13/4680
70. Gibril MBA, Shafri HZM, Shanableh A, Al-Ruzouq R, Wayayok A, Hashim SJ. Deep convolutional neural network for large-scale date palm tree mapping from UAV-based images. Remote Sensing. 2021;13(14):2787. DOI: 2072-4292/13/14/2787
71. Dehghanisanij H, Salamati N, Emami S, Emami H, Fujimaki H. An intelligent approach to improve date palm crop yield and water productivity under different irrigation and climate scenarios. Applied Water Science. 2023;13(2):56. DOI: 10.1007/s13201-022-01836-8
72. Ramakrishnam Raju SVS, Dappuri B, Ravi Kiran Varma P, Yachamaneni M, DMG V, Mishra MK. Design and implementation of smart hydroponics

farming using IoT-based AI controller with mobile application system. Chelladurai SJS, editor. Journal of Nanomaterials. 2022;2022:1-12. Available from: https://www.hindawi.com/journals/jnm/2022/4435591/

73. Canlas FQ , Al Falahi M, Nair S. IoT based date palm water management system using case-based reasoning and linear regression for trend analysis. International Journal of Advanced Computer Science and Applications. 2022;13(2):549-556

74. Dhehibi B, Ben SM, Frija A, Aw-Hassan A, El OH, Al RYM. Economic and technical evaluation of different irrigation systems for date palm farming system in the GCC countries: Case of Oman. Environment and Natural Resources Research. 2018;8(3):55

75. FAOSTAT. Food and Agriculture Organization of the United Nations. 2021. pp. 403-403. Available from: https://www.fao.org/faostat/en/#data/QCL [Accessed: 28 April 2023]

76. Ahmed Mohammed ME, Refdan Alhajhoj M, Ali-Dinar HM, Munir M. Impact of a novel water-saving subsurface irrigation system on water productivity, photosynthetic characteristics, yield, and fruit quality of date palm under arid conditions. Agronomy. 2020;10(9):1265. DOI: 2073-4395/10/9/1265

77. Mohammed M, Sallam A, Munir M, Ali-Dinar H. Effects of deficit irrigation scheduling on water use, gas exchange, yield, and fruit quality of date palm. Agronomy. 2021;11(11):2256

78. El Hadrami A, Al-Khayri JM. Socioeconomic and traditional importance of date palm. Emirates Journal of Food and Agriculture. 2012;24(5):371-385

79. Arias E, Hodder AJ, Oihabi A, Jimnez E, Hodder AJ, Oihabi A. FAO support to date palm development around the world: 70 years of activity. Emirates

Journal of Food and Agriculture. 2016;28(1):1-11. Available from: http://www.ejmanager.com/fulltextpdf.php?mno=204338

80. Ali-Dinar H, Mohammed M, Munir M. Effects of pollination interventions, plant age and source on hormonal patterns and fruit set of date palm (Phoenix dactylifera L.). Horticulturae. 2021;7(11):427. DOI: 2311-7524/7/11/427
81. Alnaim M, Mohamed M, Mohammed M, Munir M. Effects of automated irrigation systems and water regimes on soil properties, water productivity, yield and fruit quality of date palm. Agriculture. 2022;12(3):343. DOI: 2077-0472/12/3/343
82. Mohammed M, Munir M, Ghazzawy HS. Design and evaluation of a smart ex vitro acclimatization system for tissue culture plantlets. Agronomy. 2022;13(1):78
83. Statista. Leading Fruit Dates Exporters Worldwide in 2021. New York, USA: Statista Research Department. Statista Inc.; 2021. Available from: https://www.statista.com/statistics/961359/global-leading-exporters-of-dates
84. Erskine W, Moustafa AT, Osman AE, Lashine Z, Nejatian A, Badawi T, et al. Date palm in the GCC countries of the Arabian peninsula. In: Proc Regional Workshop on Date Palm Development in the Arabian Peninsula. Abu Dhabi: Icarda; 2004. pp. 29-31. Available from: http://www.icarda.org/aprp/datepalm/introduction/introduction.htm
85. Tripler E, Shani U, Mualem Y, Ben-Gal A. Long-term growth, water consumption and yield of date palm as a function of salinity. Agricultural Water Management. 2011;99(1):128-134
86. Abul-Soad AA, Jain SM, Jatoi MA. Biodiversity and conservation of date palm. In: Biodiversity and Conservation of Woody Plants, Sustainable

Development and Biodiversity. Cham, Switzerland: Springer; 2017. pp. 313-353. DOI: 10.1007/978-3-319-66426-2_12

87. Chowdhury S, Al-Zahrani M. Characterizing water resources and trends of sector wise water consumptions in Saudi Arabia. Journal of King Saud University - Engineering Sciences. 2015;27(1):68-82. Available from: https://linkinghub.elsevier.com/retrieve/pii/S1018363913000081
88. DeNicola E, Aburizaiza OS, Siddique A, Khwaja H, Carpenter DO. Climate change and water scarcity: The case of Saudi Arabia. Annals of Global Health. 2015;81(3):342-353
89. Baig MB, Alotibi Y, Straquadine GS, Alataway A. Water resources in the kingdom of Saudi Arabia: Challenges and strategies for improvement. Global Issues in Water Policy. 2020;23:135-160
90. Barth HJ. Desertification in the Eastern Province of Saudi Arabia. Journal of Arid Environments. 1999;43(4):399-410
91. Amin AA. The extent of desertification on Saudi Arabia. Environmental Geology. 2004;46(1):22-31
92. El Hassan IM. Desertification Monitoring Using Remote Sensing Technology. In: International Conference on Water Resources & Arid Environment. 2004. pp. 1-15. Available from: http://www.psipw.org/English_PDF/3_Distance/E3-3.pdf
93. Ejaz N, Elhag M, Bahrawi J, Zhang L, Gabriel HF, Rahman KU. Soil erosion modelling and accumulation using RUSLE and remote sensing techniques: Case study Wadi Baysh, Kingdom of Saudi Arabia. Sustainability. 2023;15(4):3218. DOI: 2071-1050/15/4/3218
94. Elhag M, Hidayatulloh A, Bahrawi J, Chaabani A, Budiman J. Using inconsistencies of Wadi morphometric parameters to understand patterns of

soil erosion. Arabian Journal of Geosciences. 2022;15(14):1299. Available from: https://link.springer.com/10.1007/s12517-022-10422-w

95. Salih A, Hassaballa AA, Ganawa E. Mapping desertification degree and assessing its severity in Al-Ahsa oasis, Saudi Arabia, using remote sensing-based indicators. Arabian Journal of Geosciences. 2021;14(3):192. DOI: 10.1007/s12517-021-06523-7
96. Mohammed M, Elmahmoudi A, Almolhem Y. Applications of electromagnetic induction and electrical resistivity tomography for digital monitoring and assessment of the soil: A case study of Al-Ahsa Oasis, Saudi Arabia. Applied Sciences. 2022;12(4):2067. DOI: 2076-3417/12/4/2067
97. El-Shafie HAF, Faleiro JR, Al-Abbad AH, Stoltman L, Mafra-Neto A. Bait-free attract and kill technology (hook™ RPW) to suppress red palm weevil, rhynchophorus ferrugineus (coleoptera: Curculionidae) in date palm. Florida Entomologist. 2011;94(4):774-778
98. El-Shafie HAHA, Mohammed MEME, Sallam AKAKA. Quarantine protocol against coleopteran borers in date palm offshoots using eco2fume gas. Outlooks on Pest Management. 2020;31(4):190-192
99. Mohammed M, El-shafie H, Alqahtani N, El-Shafie H, Alqahtani N. Design and validation of computerized flight-testing systems with controlled atmosphere for studying flight behavior of red palm weevil, Rhynchophorus ferrugineus (Olivier). Sensors. 2021;21(6):2112. DOI: 1424-8220/21/6/2112
100. El-Shafie HAFAF, Abdel-Banat BMAMA, Mohammed MEAEA, Al-Hajhoj MRR. Monitoring tools and sampling methods for major date palm pests. CAB Reviews: Perspectives in Agriculture, Veterinary Science, Nutrition and Natural Resources. 2019;14(022):1-11

101. Mohammed MEA, El-Shafie HA, Sallam AAA. A solar-powered heat system for management of almond moth, Cadra cautella (Lepidoptera: Pyralidae) in stored dates. Postharvest Biology and Technology. 2019;154(March):121-128
102. Mohammed MEA, El-Shafie HAF, Alhajhoj MR. Recent trends in the early detection of the invasive red palm weevil, Rhynchophorus ferrugineus (Olivier). In: Invasive Species—Introduction Pathways, Economic Impact, and Possible Management Options. London, UK: IntechOpen; 2020
103. Mohammed MEAMEA, El-Shafie HAFHAF, Al-Hajhoj MBRMBR. Design of an automated solar-powered light trap for monitoring and mass trapping of major date palm pests. Ecology, Environment and Conservation. 2018;24(1):177-185
104. El-Shafie H, Mohammed M, Alqahtani N. A preliminary study on flight characteristics of the longhorn date palm stem Borer Jebusaea hammerschmidtii (Reiche 1878) (Coleoptera: Cerambycidae) using a computerized flight mill. Agriculture. 2022;12(1):120
105. Mohammed M, El-Shafie H, Munir M. Development and validation of innovative machine learning models for predicting date palm mite infestation on fruits. Agronomy. 2023;13(2):494. DOI: 2073-4395/13/2/494
106. Sarraf M, Jemni M, Kahramanoğlu I, Artés F, Shahkoomahally S, Namsi A, et al. Commercial techniques for preserving date palm (Phoenix dactylifera) fruit quality and safety: A review. Saudi Journal of Biological Sciences. 2021;28(8):4408-4420
107. Kader AA, Hussein AM, Adel AK, Awad MH. Harvesting and postharvest handling of dates. Icarda. 2009;15(September):1-15

108. El-Habbab PMS, Al-Mulhim F, Al-Eid A, El-Saad MA, Aljassas A, Sallam SL, et al. Assessment of post-harvest loss and waste for date palms in the Kingdom of Saudi Arabia. International Journal of Environmental & Agriculture Research. 2017;3(6):01-11. Available from: http://ijoear.com/Paper-June-2017/IJOEAR-MAY-2017-7.pdf
109. Mohammed MEAMEA, El-Shafie HAFHAF, Alhajhoj MRMR. Design and efficacy evaluation of a modern automated controlled atmosphere system for pest management in stored dates. Journal of Stored Products Research. 2020;89:101719. Available from: https://www.sciencedirect.com/science/article/pii/S0022474X20304641
110. Mohammed M, Alqahtani N, El-Shafie H. Development and evaluation of an ultrasonic humidifier to control humidity in a cold storage room for postharvest quality management of dates. Food. 2021;10(5):949. DOI: 2304-8158/10/5/949
111. Almutawa AA. Date production in the Al-Hassa region, Saudi Arabia in the face of climate change. Journal of Water and Climate Change. 2022;13(7):2627-2647. Available from: https://iwaponline.com/jwcc/article/13/7/2627/89124/Date-production-in-the-Al-Hassa-region-Saudi
112. Aleid SM, Al-Khayri JM, Al-Bahrany AM. Date palm status and perspective in Saudi Arabia. In: Date Palm Genetic Resources and Utilization. Dordrecht: Springer Netherlands; 2015. pp. 49-95. DOI: 10.1007/978-94-017-9707-8_3
113. Ploennigs J, Cohn J, Stanford-Clark A. The future of IoT. IEEE Internet of Things Magazine. 2018;1(1):28-33
114. Osifeko MO, Hancke GP, Abu-Mahfouz AM. Artificial intelligence techniques for cognitive sensing in future IoT: State-of-the-art, potentials, and

challenges. Journal of Sensor and Actuator Networks. 2020;9(2):21. DOI: 2224-2708/9/2/21

115. Lin ZX, Ying ZM, Bing LQ , Chao CC, Chao TY. FinBrain: When finance meets AI 2.0. Frontiers of Information Technology & Electronic Engineering. 2019;20(7):914-924

116. Mohamed E. The relation of artificial intelligence with internet of things: A survey. Journal of Cybersecurity and Information Management. 2020;1(1):30-34

117. Mohammed M, Munir M, Aljabr A. Prediction of date fruit quality attributes during cold storage based on their electrical properties using artificial neural networks models. Food. 2022;11(11):1666. DOI: 2304-8158/11/11/1666

118. Gopalan SS, Raza A, Almobaideen W. IoT security in healthcare using AI: A survey. In: 2020 International Conference on Communications, Signal Processing, and their Applications (ICCSPA). Sharjah, United Arab Emirates: IEEE; 2021. pp. 1-6. Available from: https://ieeexplore.ieee.org/document/9385711/

119. Milić DC, Tolić IH, Peko M. Internet of things (Iot) solutions in smart transportation management. Business Logistics in Modern Management. 2020;20:331-343

120. Misra NN, Dixit Y, Al-Mallahi A, Bhullar MS, Upadhyay R, Martynenko A. IoT, big data, and artificial intelligence in agriculture and food industry. IEEE Internet of Things Journal. 2022;9(9):6305-6324

121. Prasad M, Prasad R, Kulkarni U. A decision support system for agriculture using natural language processing (ADSS). Proceedings of the International MultiConference of Engineers and Computer Science. 2008;I:19-21

122. Devi M, Dua M. ADANS: An agriculture domain question answering system using ontologies. In: 2017 International Conference on Computing, Communication and Automation (ICCCA). Greater Noida, India: IEEE; 2017. pp. 122-127. Available from: http://ieeexplore.ieee.org/document/8229784/
123. Gunawan R, Taufik I, Mulyana E, Kurahman OT, Ramdhani MA, Mahmud. Chatbot application on internet of things (IoT) to support smart urban agriculture. In: 2019 IEEE 5th International Conference on Wireless and Telematics (ICWT). Yogyakarta, Indonesia: IEEE; 2019. pp. 1-6. Available from: https://ieeexplore.ieee.org/document/8978223/
124. Yunpeng C, Jian W, Juan L. The development of deep learning based natural language processing (NLP) technology and applications in agriculture. Journal of Agricultural Big Data. 2019;1(1):38-44. Available from: https://www.cabdirect.org/cabdirect/abstract/20219909618
125. Zhang Q , Liu Y, Gong C, Chen Y, Yu H. Applications of deep learning for dense scenes analysis in agriculture: A review. Sensors. 2020;20(5):1520. DOI: 1424-8220/20/5/1520
126. Gomes JFS, Leta FR. Applications of computer vision techniques in the agriculture and food industry: A review. European Food Research and Technology. 2012;235(6):989-1000
127. Patrício DI, Rieder R. Computer vision and artificial intelligence in precision agriculture for grain crops: A systematic review. Computers and Electronics in Agriculture. 2018;153:69-81. Available from: https://linkinghub.elsevier.com/retrieve/pii/S0168169918305829
128. Li Y, Zhang Y. Application research of computer vision Technology in Automation. In: 2020 International Conference on Computer Information and

Big Data Applications (CIBDA). Guiyang, China: IEEE; 2020. pp. 374-377. Available from: https://ieeexplore.ieee.org/document/9148583/

129. Tian H, Wang T, Liu Y, Qiao X, Li Y. Computer vision technology in agricultural automation—A review. Information Processing in Agriculture. 2020;7(1):1-19

130. Tripathi MK, Maktedar DD. A role of computer vision in fruits and vegetables among various horticulture products of agriculture fields: A survey. Information Processing in Agriculture. 2020;7(2):183-203

131. Grift T, Zhang Q , Kondo N, Ting K. A review of automation and robotics for the bio-industry. Journal of Biomechatronics Engineering. 2008;1(1):37-54

132. King A. Technology: The future of agriculture. Nature. 2017;544(7651):21-23

133. Marinoudi V, Sørensen CG, Pearson S, Bochtis D. Robotics and labour in agriculture. A context consideration. Biosystems Engineering. 2019;184:111-121. Available from: https://linkinghub.elsevier.com/retrieve/pii/S1537511019303617

134. Sparrow R, Howard M. Robots in agriculture: Prospects, impacts, ethics, and policy. Precision Agriculture. 2021;22(3):818-833

135. Sivaganesan D. Design and development Ai-enabled edge computing for intelligent-Iot applications. Journal of Trends in Computer Science and Smart Technology. 2019;1(02):84-94

136. Zhang X, Cao Z, Dong W. Overview of edge computing in the agricultural internet of things: Key technologies, applications, challenges. IEEE Access. 2020;8:141748-141761. Available from: https://ieeexplore.ieee.org/document/9153160/

137. Guillén MA, Llanes A, Imbernón B, Martínez-España R, Bueno-Crespo A, Cano JC, et al. Performance evaluation of edge-computing platforms for the prediction of low temperatures in agriculture using deep learning. The Journal of Supercomputing. 2021;77(1):818-840
138. Kalyani Y, Collier R. A systematic survey on the role of cloud, fog, and edge computing combination in smart agriculture. Sensors. 2021;21(17):5922. DOI: 1424-8220/21/17/5922
139. Sharofidinov F, MSA M, Pham VD, Khakimov A, Muthanna A, Samouylov K. Agriculture Management Based on LoRa Edge Computing System. In: Distributed Computer and Communication Networks: 23rd International Conference, DCCN 2020, Moscow, Russia; September 14-18, 2020; Moscow, Russia. Vol. 2020. pp. 113-125
140. Bourechak A, Zedadra O, Kouahla MN, Guerrieri A, Seridi H, Fortino G. At the confluence of artificial intelligence and edge computing in IoT-based applications: A review and new perspectives. Sensors. 2023;23(3):1639. DOI: 1424-8220/23/3/1639

Chapter (6): Remote Sensing and Artificial Intelligence: Exploring Insect Physiological Mechanisms for Advanced Pest Management

Mohamed Abdel-Raheem[1] and Viyan Ghazi Mohammed[2]

[1]Pests & Plant Protection Department, Agricultural and Biological Research Institute, National Research Centre, 33rd El Buhouth St, (Postal code: 12622) Dokki, Giza, Egypt.

E-mail: abdelraheem_nrc@yahoo.com, abdelraheem_nrc@hotmail.com

Tel: (+2)01155527583, (+2)01009580797

[2]Plant protection Department, College of Agricultural Engineering Science, Duhok University, Kurdistan, Iraq

E-mail: viyang.mohammed@yahoo.com, Viyan.ghazi@uod.ac

Abstract:

Insects make up major part of the animal species on the planet and recent news reveals that they are also rapidly declining in number. Despite the fact that these reports are on a wide range of insect taxa and areas, the data required to determine the scope of this phenomenon is meager. The populations of insects are rather problematic to investigate, and most of the monitoring tools are time-consuming and inefficient. The development of computer vision and deep learning represents potential new solutions to this problem at the global level. Entomological observations may be well, in a continuous and noninvasive manner, conducted with cameras and other sensors over a diurnal cycle and a seasonal cycle, whereas the physical appearance of specimens can also be recorded with automated imaging in the laboratory. Deep-learning models can be trained to give estimates of insect abundance, biomass, and diversity; in addition, it can estimate variation in phenotypic traits, behavior, and interactions when trained on these data. In this respect, we relate the recent advances in deep learning and computer vision to the pressing need to develop more affordable monitoring of insects and other invertebrates. We provide the examples of sensor-based insect monitoring and show how deep-learning tools may be utilized on the significantly large datasets to obtain ecological knowledge and explain what difficulties remain on the way of the implementation of such methods in the field of entomology. To achieve this change, we give four areas of focus, which include: 1) image-based taxonomic identification validation; 2) creation of adequate training data; 3) the creation of open and curated reference databases; and 4) mechanisms to combine deep learning and molecular technology.

Key words: Remote Sensing, Artificial Intelligence, Insect Physiological Mechanisms, Pest Management

---

Introduction

Remote sensing (RS), the science of obtaining data on the surface of the earth without physical contacts has transformed the process of monitoring, analysing, and controlling of the dynamic processes of the planet. As the collection of satellite imagery, aerial photography, and ground-based sensor information has continued to grow, RS has emerged as one of the pillars in solving a broad set of scientific and societal problems, which include environmental surveillance and disaster management, city planning, and agriculture [1,2]. And yet the sheer amount, complexity, as well as diversity of data produced by modern RS technologies pose significant challenges in the data processing, analysis, and interpretation. In line with this, the artificial intelligence (AI), specifically the developments in machine

learning (ML) and deep learning (DL), have displayed remarkable abilities in the separation of meaningful patterns in multifaceted datasets. These qualities make AI the best to respond to the issues involved in the RS data analysis, such as feature extraction, image classification, object detection, and time-based analysis. [3,4].
With remote sensing, we used diverse algorithms and artificially-intelligence models that process, analyze and interpret data received by satellites, drones and other platforms. Machine-learning algorithms, including Support Vector Machines (SVM) and the Random Forests (RF) are often used to perform such tasks as land-cover classification and novelty object detection. Convolutional Neural Networks (CNNs) Deep-learning-based models are especially good at deriving spatial information in high-resolution images, when used in object detection, change detection and segmentation. More sophisticated algorithms like Principal Component Analysis (PCA) are useful in dimensionality reduction, whereas clustering algorithms like the K -Means and Density -Based Spatial Clustering of Applications with Noise (DBSCAN) are useful in unsupervised classification [5,6].
Remote sensing has interacted with artificial intelligence, creating new opportunities to enhance the innovation of solutions thereby making them more accurate, efficient, and automated in different domains of application. Moreover, it can be reinforced with the fact that these developments are facilitated by an expanding ecosystem of available datasets, ready-to-use models, and computing infrastructure that has made the ingestion of AI approaches widespread in the remote-sensing community. Although such encouraging advances are present, remote sensing and artificial intelligence also pose new challenges. Matters like the ability to generalize models in different geographical areas, the interpretability of AI models, and ethical concerns of automated decision-making are factors that should be taken into account. In addition, there is a key bottleneck of bridging the

information gap between domain knowledge in remote sensing and technical skills in artificial intelligence to facilitate wider adoption and success. [**7,8**]. Recent scientific observations have shown that there is a mass extinction of species. Nevertheless, information regarding species diversity and abundance changes is plagued by large taxonomic, spatial, and temporal biases and gaps. This is very strong especially in insects, which constitute the huge majority of animal species. A significant reason behind these data deficits when it comes to insects and other invertebrates is the fact that the means by which the study and the population tendencies of the species can be researched and monitored are archaic and ineffective. However, recent literature has revealed frightening statistics of the loss of insect resources and abundance. In order to go deeper into understanding the degree and reasons behind such changes, there is a need of effective, strict, and dependable means of studying and tracking insects. The population trends of insects are already being produced in the form of data in the context of current biomonitoring programmes. Legislative terrestrial biomonitoring (e.g. as part of the European Union Habitats Directive) however is limited to a very limited number of individual insect species, e.g. rare butterflies and beetles, due to the impossibility or prohibitive expense of monitoring most insect taxa. Invertebrates are widely used in ecological status assessment in the present-day legislative aquatic monitoring, such as through the Clean Water Act of the US, the Water Framework Directive of the EU, and the Marine Strategy Framework Directive of the EU. Nevertheless, the spatiotemporal and taxonomic scale and resolution of current biomonitoring programmes is crude and lacks the detail of the status of very large proportions of the insect populations. In the future, molecular methods including barcoding and metabarcoding of DNA might be useful when it comes to insect monitoring using field-collected samples; however, currently, high-throughput approaches cannot rely on the ability to estimate abundance, which creates a serious gap in the methodology. The art of deep learning, computer vision, and image processing has now reached a stage where it can assist - or even substitute - manual observation on the spot as well as in standard laboratory functions of process to samples. Camera traps are image based observational techniques that have experienced a rapid development and are used to monitor vertebrates in the last decade. Equivalent methods that involve the use of cameras and other sensors to study the variety and richness of insects are being implemented. Nonetheless, in other fields, deep learning is slowly being used in invertebrate monitoring and biodiversity studies, despite a significant level of attention given to the latter. Deep learning models are trained to extract features of a dataset, and they do not require manual features extraction, instead, learning features through repeated training on a set of examples. In such a sense, deep learning is qualitatively different to the traditional operational statistical approaches to prediction. By use of deep learning models that are specially trained to process images (also known as convolutional neural networks, or CNNs), features in images or objects in them can be extracted, and can be trained to distinguish between them. The automatic identification and categorization of insects in video or time lapse images using trained CNNs have very high potential in terms of monitoring insects. These approaches will provide new thrilling possibilities of understanding insect ecology and monitoring as they continue to be advanced.[13].

This chapter will argue that deep learning and computer vision can be utilized in the development of new systems with high throughput to detect, enumerate, classify and discover species and derive functional characteristics such as biomass, useful in biomonitoring applications. These techniques can contribute to the long-term dilemmas in the study of the ecology and biodiversity as well as can help solve the urgent problems related to monitoring the insect population. There are three objectives of this article. To begin with we will show sensor-based systems to observe invertebrates in the natural environment, as well as to study them in the laboratory using specimens. Our solutions are concerned with those that already make use of or have the potential to make use of deep-learning models to process the huge masses of involved data. And lastly, we summarize and comment on the four main issues which are the future of such solutions to monitor invertebrates, their ecology as well as biodiversity studies [14, 15]. Monitoring of Insects by Sensors. There is extensive use of sensors in ecology to collect peripheral information like light, precipitation and temperature. However, only in recent times have solutions with sensor-based monitoring insects and other invertebrates in the natural environment been offered. The agricultural research is mainly concerned with the innovation and development as it aims to predict abundance and occurrence of the economically valuable beneficial and pest insect species, to screen natural products against invasive insect species more efficiently, or to monitor disease vectors like mosquitoes. Cameras, radars and microphones are the most widely used. This sensor based monitoring is bound to create datasets that are many orders of magnitude larger than any previously studied in ecology (i.e. big data), and thus require effective solutions to extract meaningful biological information. Deep learning can become an important instrument in this respect. Our examples of image-based methods of tracking insects appear below, and we believe that they represent the most promising in terms of integrating deep learning. Also, we explain strategies that utilize alternative forms of sensors, where the integration with deep learning remains unsophisticated but may also hold some applicability to the detection and classification of entomological data. We also outline the current attempts to digitalise natural history collections, which may produce useful reference information to train and test deep-learning models. [16-18].

In Situ Monitoring Image-Based Solutions

Cameras and deep learning have been used in some case studies to detect single species, such as the olive fruit fly, the Bactrocera oleae, or more generally pests. In such cases, the identification of pests is based on the images of insects detected by the use of a device like McPhail traps or pheromone-lure traps having adhesive liners. A microcomputer captures the images and sends them to a remote server where they are analyzed. Other solutions combine a digital camera and microprocessor that can count trapped specimens in real time through object detection based on deep learning models. Under the two conditions, deep-learning networks are trained to identify and list people. However, not many field studies on invertebrate biodiversity have used the deep-learning models. The first experiments used feature vectors based on single-view images with relative accuracy of 35 moth species, or based mostly on coarse taxonomic resolution. Recently, we have demonstrated that our specially designed time-lapse cameras can capture image data, which can be used by a deep-learning model to predict local spatial, diurnal, and seasonal processes of honeybees and other insects that visit flowers (Fig. 1). Time-lapse photography is not prone to observer bias as direct observation, and data collection may take a full diurnal and even seasonal cycle. Cameras can be either baited like the traditional light and pheromone traps or placed above temporary natural resources like flowers, fruit, dung, fungi or carrion. Bjerge et al. suggest the use of an automated light trap to observe the abundance of moth and other insects attracted to light. The system uses power supplied by a solar panel which can be installed in remote places (Fig. 2). Over the long term, an entirely Internet-of-Things-enabled hardware will enable the use of classification algorithms to be run entirely on camera units and deliver entirely autonomous field systems to track insects and send back detection and classification data on real-time to the user or online portals.[18-23].

Fig. 1.

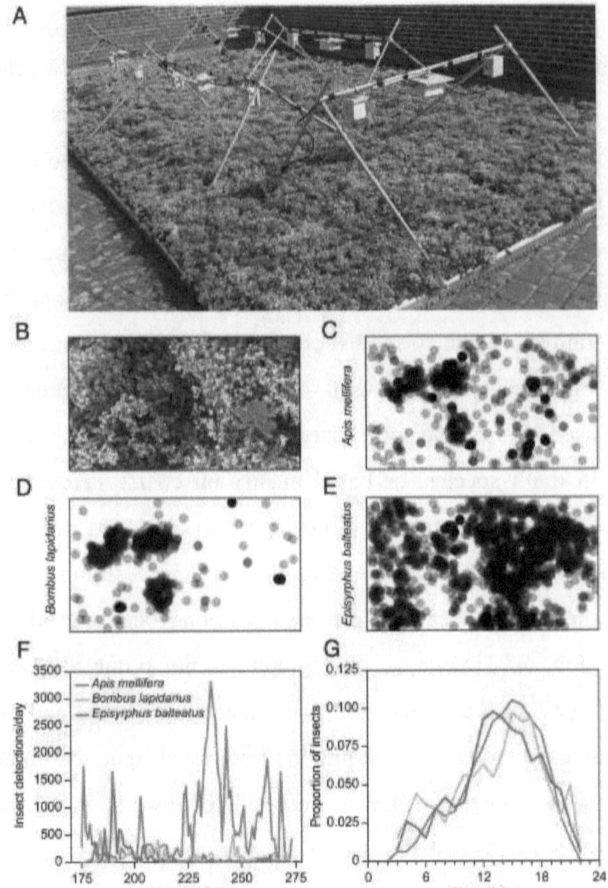

A camera trap monitoring insects visiting flowers was designed and used in this study to take photographs at intervals. The system is composed of two web cameras connected to a control unit with Raspberry Pi computer and hard drive. In our experiment, ten camera traps were fixed to the custom made steel rod supports, which were placed 30⁻ 1 cm above a green roof made of the Sedum species. The pictures were taken after 30 s during the whole flowering period. We used a convolutional neural network (Yolo3) to train and, throughout the entire growing season, were able to detect over 100,000 instances of pollinators. (B) There is an image of one of the cameras, which depicts a camera view consisting of multiple flowering plant species. The European honeybee (Apis mellifera), red-tailed bumblebee (Bombus lapidarius), and marmalade hoverfly (Episyrphus balteatus) are three typical flower-visiting insects with a significant difference in location of their initial detections by three species. In the ten image series, the deep-learning model was able to identify finer variation in (F) seasonal and (G) diurnal variation in the appearance frequency of the same three species. Altered with permission of the reference.[24].

Fig. 2.

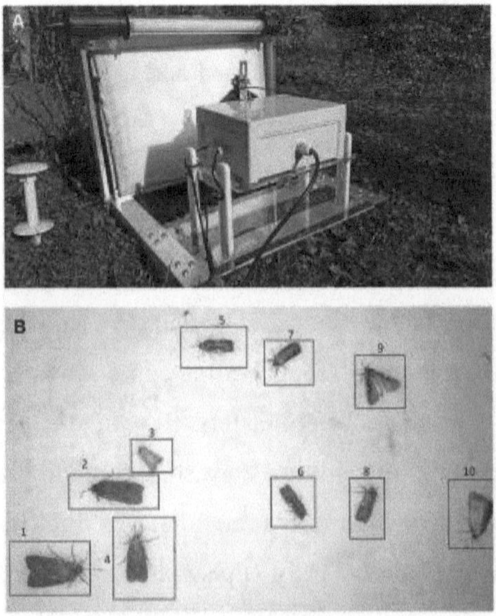

(A) A light trap fitted with an onboard computer vision system was developed in order to monitor nocturnal moth species automatically. The trap contains three different sources of lights; a fluorescent tube which attracts moths, a light table with white sheet which gives the trap diffuse background light to rest insects and a light ring to light up the specimens. The moths are attracted to the system and automatically captured pictures are taken following the motion detector. The materials used in the construction were standard ones such as a high-resolution universal serial bus web camera and a Raspberry Pi computer. (B) A computer vision algorithm has been suggested which tracks and counts individual moths. A specialised version of a convolutional neural network was trained to estimate and categorise eight different species of moths. In this given photograph, ten (1-10) people were automatically identified. The algorithm has the ability to run on the onboard computer and through which the system can process and transmit species data through a modem to a server. A battery and a solar panel provide power, which generates the possibility of working off-grid. The following with permission of the reference [25]. In Situ Monitoring Radar, Acoustic, and Other Solution. Radar technology has led to the ability to study insects in heights of ranges that could not be studied using conventional techniques, especially both non-migratory and migratory insects. Information based on conventional weather radar networks may be used to give the information of continental level, whilst specialised radar tools like vertical-looking radars (VLRs) can be used to provide information at a smaller scale locally [26-28]. VLRs are capable of determining the biomass and body shape of the object that has been detected, the direction of flight, speed and body orientation can be obtained with the help of the returned radar signal. However, VLR data provides little understanding of community structure and conclusive species identification is yet to be achieved by use of aerial trapping. Low altitude scanners (Harmonic scanning radars) are capable of detecting insects at low altitude (more than a few hundred metres) that have a radar transponder on them and must be in the clear line of view. The overall implication of applying radar technology in entomology is valuable information in insect monitoring, such as on the scale of the biomass flux caused by insect migrations but it requires confirmation with additional techniques. Bioacoustics is an old scientific field and acoustic signals have been used extensively in the ecological field. Although the majority of bioacoustic methods are used with birds and mammals in mind, entomological monitoring can be beneficial through the application of bioacoustic techniques. Research has shown that classification of groups of beings like grasshoppers and bees is possible through machine learning in the context of audio metabolism. It has been suggested that using pseudo- acoustic optical sensors instead of real acoustic sensors is a more promising technology since the signal to noise ratio of such systems is much higher. However, the deep-learning techniques can be a useful technology in acoustic entomological surveillance.[29- 33].

The other sensor technology is used to automate the process of recording insect activity or body mass, but the further processing of the information gathered using deep learning techniques is not always taken into account. In another recent study, photodiodes and infrared LEDs were formed into a sensor ring that was used to detect both large and small arthropods, such as pollinators and pests, with a detection rate of 95 per cent live microarthropods of three different species with a size range of 0.5 to 1.1 mm. The Edapholog is a low power monitoring system that was developed to detect microarthropods in soils. Its probe and sensing mechanism depends on the measuring of the variation in intensity of the infrared light; it counts the organisms falling in the traps and measures the size of their bodies. The probe is connected through radio signal to a logging device that forwards the information to a server where real time monitoring will be done. On the same note, some other researchers have also enhanced traditional low-cost trapping methods with the use of optoelectronic sensors and wireless communication that could be used in real-time monitoring and reporting [34-38]. Since there is no intuitive set of images generated by such sensors to validate, a sufficient amount of validated training data in deep learning models might be difficult to obtain, but such models still can be useful. Digitization of Specimen and Natural History Collection. There are several efforts to digitize natural history collections due to a wide range of reasons, such as the opportunity to apply deep learning methods. The advantages and need to digitize natural science collections have led to the creation of the Distributed System of Scientific Collections Research Infrastructure (DISSCo RI; https://www.dissco.eu/). The aim of DISSCo, RI is to digitize all European natural science resources on the basis of shared curation and access policies and practices.

The majority of databases in existence today include single-view images of pinned specimens; datasets retrieved on insect specimens measured with a series of sensors, three-dimensional models, and databases of living insect specimens are only just beginning. The latter may be especially applicable to deep learning models. Moreover, she has a good store of entomological information in the herbarium specimens in the shape of herbivory signatures. Herbarium collection data has been shown to be standardized and suitable in extracting the data on herbivory using machine learning methods. The methods that will be used to digitize will help in bringing such good databases faster through automation techniques. BIODISCOVER machine is an automated digitalization device of liquid-preserved specimens including the majority of insects collected in the field [3943]. It involves four automated processes: 1) picking of individual insects into bins of bulk samples; 2) capturing the specimen by high-speed imaging of the specimen in various angles; 3) storing the data of the captured specimen in a highly efficient way to be used in deep learning algorithms and further analysis; and 4) sorting the specimen based on their size, taxonomy or rarity to be used in additional molecular processing. The introduction of these tools to help process bulk insect samples based on large-scale inventories and monitoring studies would be able to generate population and community information very quickly and nondestructively. The digitization should also pay attention to the manner in which the images of individual specimens can be used to create deep learning models to be utilized in in-situ monitoring.

Fig. 3.

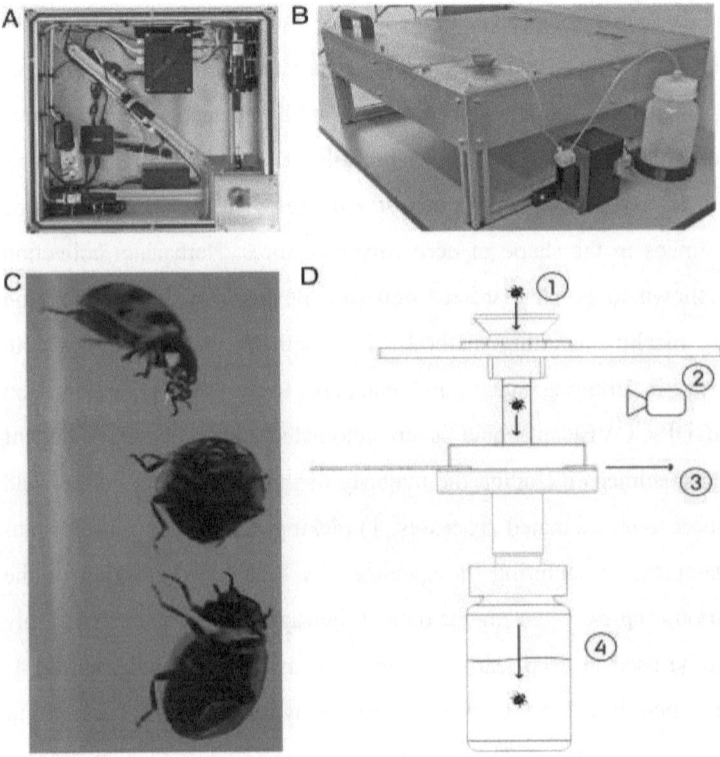

The BIODISCOVER machine has the ability to automate sample sorting, species identification, and biomass estimation of invertebrates samples. (A) An ethical system includes; the ethanol filled spectroscopic cuvette, a very strong and controllable source of light, and two cameras that are capable of capturing images at 50 frames per second. The device has a refilling pump to fill the spectroscopic cuvette and (B) is housed in a light-proof aluminum box. (C) The cameras see each specimen in two angles when it is inserted into the ethanol-filled cuvette, and geometric characteristics related to size and biomass calculated automatically. (D)

Two cameras record the image of the specimen as it passes through the ethanol. The system includes an inbuilt flushing mechanism to control the type of specifications that should be kept together to allow some form of storing or analysis to be done. The findings of the first dataset including images of 598 specimens of 12 known-identity species were very promising with an achieved classification rate of 98.0. Taken after being licensed under CC BY 4.0. [44]. Deep Learning Potential uses in Entomology. The massive amounts of data that sensor-based insect surveillance produces in the above context demand effective remedies to these problems of transforming the data into biologically significant forms. In this regard, preliminary findings suggest that deep learning is a useful tool and can in addition promote the purchase of new forms of data. Deep-learning software (e.g., in ecological applications) is most often built on top of open-source Python packages and frameworks (e.g., TensorFlow, Keras, PyTorch, and Scikit-learn) [34], and could be prototype code, typically made publicly available (e.g., on https://github.com/). The recent developments in other areas connected with object detection and fine-grained classification are therefore available in the entomological researches. Therefore the deep-learning toolkit is already available to entomologists, though some might need modification to entomological use. We briefly describe the transformative capability of deep learning regarding entomological data, stored in images, below, divided around four main applications [35]. Surveillance and Tracing of persons on the ground. The monitoring of insect abundance and diversity through imagery will become a universal practice in the world as the nations strive to understand the magnitude of the world insect loss, and what can be done to mitigate the situation. Online portals like https://www.inaturalist.org/ and https://observation.org/ and the corresponding smartphone apps have recently made the identification of individual insects easier.

They are systems that offer instant candidate species upon the user uploading pictures of observed insect species with deep-learning models. Even though these portals provide powerful means of quick biodiversity data generation, they are used not to layer ecology but to opportunistically record the occurrence of the species. However, these systems may be modified to serve as a tool of monitoring. Indicatively, a system of time-lapse cameras may produce high-temporal and high-spatial-resolution image data of tracking certain insect species. In some situations, it may be possible to detect the persons based on such data with simple changes in the existing designs and in some situations, the specific designed features might be required. The solution should be the combination of image data with acoustic or behavioural data, which are harder to detect and identify taxa. Moreover, object-detection models have the potential of identifying and categorizing people besides the location of the specific individuals in an image. Such models may be used on time-lapse and video data to monitor the location of individual insects in place during time. This would provide a further useful layer of data, which can be based on observations in the form of images. As an example, the observed microclimatic variation can be attributed to the movement speed of individual insects, and more realistic curves of thermal performance can be drawn and compared to the conventional thermal performance derived in the laboratory. Nevertheless, tracing insects within the wild area is one of the most complicated tasks nowadays because of such problems as contaminated scenes, and different lighting. In computer vision, they are known as 1)detection-based online multiple object tracking and 2)they operate under a series of assumptions. They assume that there must be accurate initialisation (initialisation) of the objects to be tracked in a scene; there must be good visual discrimination of the various tracked object; and the tracked objects must move smoothly with their velocity and acceleration (72). The minute visual variation of a single insect as well as the hiding behaviour contravene the above assumptions. In addition, state-of-the-art deep-learning models today typically have millions of learned parameters, and can only execute in real time with low-resolution video, which limits the visual discrimination of the objects of interest in the scene. The use of nonlinear motion models and the creation of compact or compressed deep-learning models are those issues that may help to resolve these challenges. [45-50]

The resolution of the tracking individual insects issue can help to create a personalized ecological model that has far-reaching effects in the research specialties of population, behavioural and thermal ecology and even conservation biology. In addition, depending on the new developments in low-price powerful graphical processing units and specific artificial intelligence processors that can be used in autonomous and embedded systems (such as NVIDIA Jetson Nano, Google Coral Edge TPU, and the Intel AI USB stick), it is not impossible to imagine that it will become possible to detect, track, and decode insect behaviour in real-time and the data it generates will be sent back to the user. Uncovering Species Interactions. The interactions between species are necessarily critical to the operations of ecosystems, but due to their temporality and speed, the effects of disruptions on the ecological operation are hard to measure. The quantification of species interactions is uniquely determined by high-temporal-resolution image based monitoring of consumers and resources. An example can be insect flowers, defoliation by herbivores, and predation episodes that can be constantly monitored during full growing seasons with fixed positions cameras. Recording of such interactions should be recorded at scales of interaction i.e. by watching interacting people at time intervals in seconds to minutes and better still the recording should be carried out over seasonal and/or multi-annual periods. Our initial findings indicate a promising prospect to capture time-lapse videos of plant-insect interactions with time-lapse cameras and deep-learning models [51-53] (Fig. 1). Taxonomic Identification The problem of taxonomic identification can be considered a deep-learning classification one. The image of a specimen can be classified by deep-learning-based methods with higher accuracy that is close to the methods of human experts to identify insects. Even CNN classification can have morphologically important features visualised by applications of gradient-weighted class-activation mapping. The accuracy of the classification tends to be reduced by the time insects are observed live in the wild; however, when it cannot be assured that the taxonomic rank used to classify an insect is going to be correct at the species rank, it may still be possible to obtain a high level of confidence in classifying insects at a higher level of taxonomic resolution. CNNs have achieved impressive results in the recent years. Large image datasets, e.g., 1,000 classes ImageNet dataset, can be classified by them with high accuracy and speed [54-56]. The best CNNs have classification accuracy similar to botanical experts even with images of more than 10,000 species of plants. This performance is possible today only when using very large volumes of training data, and can be further enhanced, and this is due to recent encouraging results in the distributed training of deep neural networks and federated learning [57]. It is typical that ecological communities often have a high percentage of comparatively rare species. This frequently leads to extremely skewed datasets, and the amount of specimen of the rarest species might not be big enough to train neural networks. Therefore, enhancing the level of algorithms and methods of better recognition of rare classes is among the main challenges of deep-learning-based taxonomic identification. Resolutions to this problem can be based on the resampling of classes and cost-sensitive learning or on multiset feature learning. Class resampling aims to be able to balance its classes by undersampling the larger classes and/or oversampling the smaller classes, whereas cost-sensitive training gives a greater loss on the smaller classes. Under multiset feature learning, larger classes are divided into smaller subsets and combined with smaller classes to create separate training sets. All these approaches attempt to acquire characteristics that better differentiate the smaller classes. The performance of species identification can be different with some species well identified in most of the instances to other species that are usually hard to identify. The volume of training data is usually a determining factor to successful identification, recent results of images of about 65,000 specimens of carabid beetles collection at the Natural History Museum, London, indicate that imbalances in identification performance are not always associated with the effectiveness of a species representation in the training data. The large datasets have not been completely studied, and thus more efforts are necessary [58-65]. Another related issue is presented by those species that are not found at all in the reference database, where the deep-learning models are trained. Detection of such species entails methods that are devised in multiple-class novelty/anomaly detection or open-set/world recognition. In one of the recent surveys, a number of open-set recognition approaches were introduced, and two major ones are discriminative and generative. The discriminative models use the classic machine learning algorithms or deep neural network with further mechanisms to identify the outliers, whereas the primary concept of the generative models is to use the positive or negative samples to train. Nevertheless, the existing procedures are usually used on comparatively low scales of data and can not be easily scaled to the amount of classes. The classes in insect datasets are often numerous and their distribution is very fine-grained with the difference between phenotypes in species possibly minute but intraspecific variation potentially extensive. Open-set recognition methods are particularly difficult to do with such datasets. Although this challenge will be incredibly hard to surmount with the methods of phenotype-based identification, a combination of image-based deep learning and DNA barcoding tools can be used to address the issue. [66-68].

The pest management precision agriculture requires constant monitoring of the target pest population, along with constant evaluation of the environmental factors, including temperature and humidity. The existing methodology is based on the principle of pest entrapment into pheromone-based adhesive traps that can be commercially purchased and can attract insects in extremely low concentrations as shown in Figure 4. Practically, every insect caught is determined by the farmer either by daily inspections in the field or by crude wireless camera devices in an attempt to determine whether the insect is a codling moth. Such an approach is not as sophisticated as an Internet-of-Things solution and in itself is constrained due to multiple factors: (1) the possibility of human error in the process of identifying and listing pests, (2) the necessity of a skilled employee, and (3) inefficiencies caused by the fact that traps may be inspected only after the delay. Therefore, the traps are frequently found out in time when they will not be effective exhaustively. In a bid to overcome these inefficiencies, the current paper is an innovation of a cost-effective pest management system, which can be implemented straight in the field to identify, detect and quantify pests. The designed system has an accuracy of 97.50 in detection, recognition and counting with the help of machine-learning and deep-learning algorithms. In addition, the proposed solution will include end-to-end solution that will be based on a secure architecture to guarantee safety and integrity of transmitted information.[69].

Fig. 4

Traditional pests trap where pests are monitored by trapping them in pheromone-based glue traps and include manual scouting.

The current research has been inspired by various reasons and factors. They can be outlined in the following: Established agricultural businesses, governmental departments, and agencies are also increasingly interested in coming up with secure tele-farming systems to use at large scale. These systems will be self-governing and artificial intelligence. The research community is made ineffective by the lack of publicly available, well-organized sets of pests. This kind of data would be the base of the machine-learned models that would help to quickly identify, locate, classify, and manage pest infestations. According to literature, the available records of data on the pests are limited and fragmented. Therefore, data collection and real-time analysis are necessary to provide the implementation of optimal practices and timely execution of control measures. The implementation of an IoT-enabled smart trap with the high-security level will increase awareness regarding the most dangerous population tendencies of pests in Pakistan. The increased knowledge of pest taxa active through the seasons and regions of the country will be useful to farmers and other interested parties. The findings of the paper can be summarized as follows: The analysis of the latest and the most topical publications on the topic of pest management, along with the use of machine-learning tools in the field of precision agriculture, are provided. A low-cost, smart, IoT-based pest detection platform design to enable integrated pest management and condition monitoring of crop-fields is described. These platforms are also useful to farmers in the actual field conditions through optimisation of inputs and treatments. The compiling of a pest-related dataset to the wider research fraternity is done. The process of creating a machine-learning model to manage pests is outlined. This model which will be trained on the dataset obtained is likely to examine and categorize pests in real time thereby enhancing the efficiency with which farmers and authorities will take appropriate actions. A comparative and experimental study of the proposed model is made. In contrast to the current literature, which uses either conventional traps or smaller machine-learning models, and often does not have a structured pest dataset, this paper combines an IoT-enabled pest detection system with the powerful convolutional neural network (CNN) models and You Only Look Once version 8 (YOLOv8) framework, therefore, considerably improving the accuracy of pest-detection processes in the field. Besides solving the problem of lack of structured pest data, the proposed solution can provide an affordable and scalable tool to farmers that enhances pest-management methods and helps to utilise agricultural inputs more productively. Furthermore, the study is distinguished by the development of the publicly available data which will facilitate the research and development of the precision-agriculture industry. They are IoT, machine-learning, and state-of-the-art frameworks, mechanisms of creating a smart-trap framework, and other data-communication techniques. In article [70], the authors explain that smart traps based on visual imaging are becoming increasingly popular and how the demand of such traps is growing day by day. These traps can now be commercially obtained and there are a number of companies that make intelligent traps that are becoming more efficient in their manufacture. The authors also review the literature of many other researchers to obtain an idea about the nature of the pests, how the farmers have been fighting with the pests and how the smart traps have been improved over the years. They also answer an issue of either mobile or stationary traps are more effective on insects though they were unable to establish a definite preference on the issue because of the cost pain related to traps mobility on fields. The authors were able to show in simulations that a single trap on a circular orbit had a high capture rate at the trap speed. The main aim of the paper, therefore, was to help in designing and developing of a sophisticated smart trap that would be beneficial not only to the trap manufacturers but also the pest-management practitioners. The companies that have developed traps can make informed decisions to develop a trap that will capture, count and promptly give signals via alarm using all the details of the trap, this is because they have a complete understanding of the weather conditions, the amount of crops to be loaded onto the trap and all other relevant information that will help the trap developers develop a trap that will capture, count and give an immediate signal, using the information. Finally, the authors conclude that given different viable methods, traps might become intelligent enough in order to help ensure less harmful and cheaper food supply.

In[71], the authors looked at processing of human learning and memory patterns to maintain high results on new and legacy tasks based on the human ability to accumulate information and reduce forgetting. They suggested a continual classification methodology based on the Artificial Neural Network (ANN) which is founded on memory storage and retrieval, and their results display two major benefits, namely, fewer data requirements and increased flexibility. Such advantages are obtained through the combination of Convolutional Neural Networks (CNNs) and Generative Adversarial Networks (GANs). To solve the problem of data scarcity, the metric-learning model based on CNNs reduces the number of data, reducing the challenges in images and the number of annotations, which is important to simplify the process of classification and pattern recognition. Enhanced flexibility is achieved due to the integrated CNN-GAN design as discussed in detail in the sources [7274]. The GAN component performed the extraction of salient information of legacy tasks and the synthesis of abstracted images as a memory store of the upcoming tasks. As a result, the refined continual model, which was used in the classification of pests and plants, reduced the forgetting issue and attained acceptable performance and high effectiveness. The authors believe that the suggested approach can be useful in various areas, starting with the agricultural sectors and other areas where sophisticated intelligent technologies can be used like robotics, and intelligent traps to harvest fruits and to detect diseases and pests [75]. The main weaknesses of the study are in the fact that GAN component is not stable and has a complicated training process, unlike the CNN model that is simpler to implement. The weaknesses of the GAN in terms of instability and difficulties in training can undermine the overall model performance. The authors presented DeepPest, a cascaded convolutional neural network structure, in[76, 77] that is useful in detecting small objects. DeepPest is a two-phase mobile vision system that leverages on the ample multi-species pest data to automatically recognize and detect pests in-situ. The initial phase uses context-sensitive attention network incorporating a multi-scale information to obtain primary classification of crop categories. The second step refers to training a multi-projected pest-detecting architecture with such crop-categorised images to boost the effectiveness of the in-field pest-detecting via super-resolved feature-detection. Although this method is a good way of detecting small pests correctly, it might fail to detect large pests as the contextual information has little impact. In[28], the authors introduced a crop-pest detection algorithm based on multiple deep convolutional neural networks that will accurately detect ten common crop-pest species. It has introduced the main results: a validated and shared crop-pest dataset, a fined tuned GoogLeNet model suggested to classify pests in agricultural settings better, and a 6.22 per cent outperformance of the state-of-the-art approach. In[79] the authors used a three model approach of identifying and classifying insects. The former model was an invariant local-feature model learnt with the Hessian-Affine detector. The second model was a global-feature model based on k-means clustering and Euclidean distance distance to determine the hue value of each point of two clusters. The third model was a hierarchical combination with two levels, whereby the first level was the selection of the best-performing classifier of global features. When a threshold value of classification probability was reached, the samples were moved to the second level where the most successful classifier of local characteristics was chosen. The results obtained by using the third method had a maximum accuracy of 96.9 per cent on the first level using Normal Density - Based Linear Classification and the accuracy of 71.3 per cent on the second using the Nearest Mean Classifier. With a combination of all these approaches, this resulted in a 96.9 per cent accurate classification of lab-acquired images. When the same third method was implemented in the field conditions, the accuracy of the correct classification was 65.5. This study involved the identification of white-fly insects in traps of adhesive surfaces.[80].

The image was first converted to gray scale and then the Sobel edge detector filter was applied to extract the image's magnitude gradient. After that, a binary image was created using the Otsu algorithm. To remove isolated spots from the image, a 3 3 median filter was applied to the whole image. The segmented objects' area, center of mass, eccentricity, and solidity were extracted at the end of the segmentation process. These features were used to group the objects into clusters, one of which was made up of features that represent white flies. The average accuracy was 97%.

In[81], the goal was to identify white flies and aphids in adhesive traps as well. The samples used were 21 21 pixel images that fall into three categories: background images of the trap (that do not include any insects), aphid images, and white fly images. Three different feature extraction techniques i.e., colors, Gabor filters, and Pyramidal Histogram of Gradients (PHOG) were employed in the identification process. The final classification was done with Support Vector Machine (SVM), using Radial Basis Function kernel (RBF) for PHOG and color features, while linear kernel for Gabor filters. As a result, 91.8% of aphids and 98.5% of white flies were detected. In contrast to the research works cited, the proposed model does not rely on the explicit feature extraction from the image processing. Instead, it utilizes the entire image to identify the flies. This approach not only simplifies the model but also enhances its adaptability to different kinds of insects. An IoT-enabled pest detection and classification model employing YOLOv3 is presented. The CNN Long Short-Term Memory Networks (CNLSTM) classifier receives the images after they have been extracted by CNN. In addition, hyper parameter optimization is done using the Adapti The picture was initially turned into gray scale and then Sobel edge detector filter was used to get its magnitude gradient. Then a binary image was generated using Otsu algorithm. To remove the isolated spots, the median filter of 3 x 3 was used on the whole image. The solidity and eccentricity of the objects segmented, as well as their area and centre of mass, were obtained at the end of the segmentation process. These features were subsequently used to group the objects with one group consisting of features that are representative of white flies. The mean of the accuracy was 97%. In [81], the aim was to detect both the white flies and aphids in adhesive traps. The sample was in the form of 21 21 pixel and it was divided into three categories: the background images of the trap which have no insects, the image of aphids and the image of white flies. The three different feature-extraction methods that were used in the identification procedure include colour, Gabor filters and Pyramidal Histogram of Gradients (PHOG). The last classification was done using a Support Vector Machine (SVM); PHOG and colour features were classified using RBF kernel and Gabor-filter features were classified using a linear kernel. Consequently, 91.8 per cent of aphids and 98.5 per cent of white flies were properly identified. In contrast to the mentioned works, the model suggested does not use explicit feature extraction based on the processing of images. It instead applies the whole image to find out the insects, thus making the model simple and flexible to most types of insects. A model of pest detection and classification using the YOLOv3 and connected to the IoT is provided. Convolutional Neural Network

Long Short-term Memory Network (CNLSTM) classifier gets the images when they are extracted using CNN. Further, hyper parameter optimization is done through Adaptive Honey Badger Algorithm (AHBA) [79-83]. In [84], the authors proposed a new method of detecting two species of fruit flies: the Ceratitis capitata and the Grapholita molesta that can be used to track insect population in intelligent field traps. The pests were detected with the help of digital image-processing techniques and they were classified with the help of Artificial Neural Networks (ANN) into the correct species. Different CNN architectures of various configurations were trained on a field-captured image dataset. The maximum precision of the study was 93.55 per cent. of the species, Ceratitis capitata and 91.28 per cent. of the species, Grapholita molesta. In addition, ResNet-18 and SqueezeNet produced the shortest classification time. Based on the smart-trap network, the Figure 5 provides the proposed system architecture. The architecture includes the pheromone dispensers, Raspberry, Pi camera, DC fan (to clean the trap during need) and LED bulb (to provide light) installed in the smart trap. Pheres are insect attractants (liquid chemicals) and used as insect attractants, the current mixture would have 85 percent methyl eugenol, 10 percent sugar, and 5 percent DDVP. This blend is sprayed on cotton to work as a male-sex attractant. The pheromones have a range of fragrances and the range is 500m and is relative to the direction of the wind. When insects are attracted, they will be trapped by the sticky surface and the Pi camera will automatically take a picture and the image is stored on the local storage of the Raspberry Pi. Convolutional neural networks are used to analyse the captured images and identify a pest fruit fly. CNN and YOLOv8 are used to assess the effectiveness of the proposed system to see which one provides better results. The pest images are stored in the cloud server (Firebase storage) and the numbers of the pests in each image are counted. These numbers, the date, and time of capture are sent to the real-time Firebase database. The smart trap is also waterproof to resist the sun and rain. It has a 4G USB dongle that transmits images of the pest and other calculated data to the cloud server. The electric power is provided to everything by a solar panel. As a result the suggested system is efficient in monitoring and capturing pests. There is also the use of synthetic pheromones which are used to lure the pests to the trap and hence successfully control them.

Fig. 5

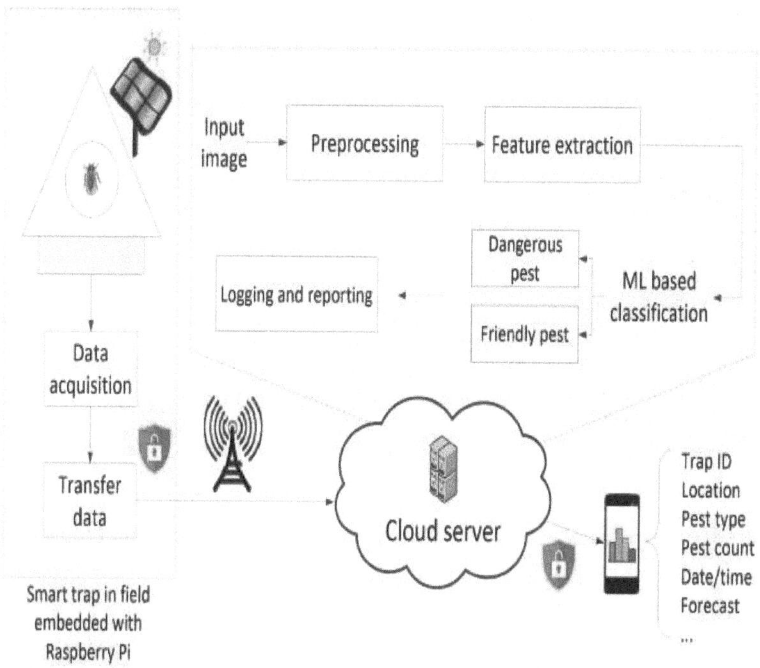

Intelligent smart trap architecture consists of pheromones, a Raspberry Pi, a Pi camera, a DC fan, and an LED bulb present inside a smart trap for monitoring and capturing the pests.

The complete system is shown in Fig. 6, which performs Machine Learning-based categorization at the edge device and transmits the findings to the stakeholders via the cloud, i.e., the Google Firebase real-time database.

Fig. 6

Intelligent smart trap system which performs Machine Learning-based pest detection at edge device and transmits findings to stakeholders via the cloud and includes a solar-powered IoT-enabled smart insect trap

New AI-Based Pest and Disease Detection and Control. As Figure 7 demonstrates, the workflow starts with the data acquisition, including RGB and multispectral images, which were taken with the help of unmanned aerial vehicles, satellite images, and the textual information, which was extracted to agricultural databases. It is these heterogeneous inputs that form the basis on which further analysis can be done. At the second phase, visual, environmental, and textual data are merged with the use of deep learning and multimodal fusion methodologies, such as convolutional neural networks, recurrent neural networks, and transformer architectures, and small-sample augmentation methods are used to prevent data shortage. Intelligent control operations, which include pest and disease detection, prediction and early warning, precision sprayers, and biological control are direct results of this stage. Lastly, the figure outlines two levels of potential directions (1) research-level innovations, as noted in the left panel, including causal reasoning models, digital-twin virtual farms, multimodal integration systems and eco-regulation modelling; and (2) system-level development that is shown in the bottom panel and emphasizes the use of digital-twin systems in sustained intelligent agricultural management. Such schematic shows how the data acquisition has led to intelligent control and outlines both theoretical and pragmatic paths in the future research.

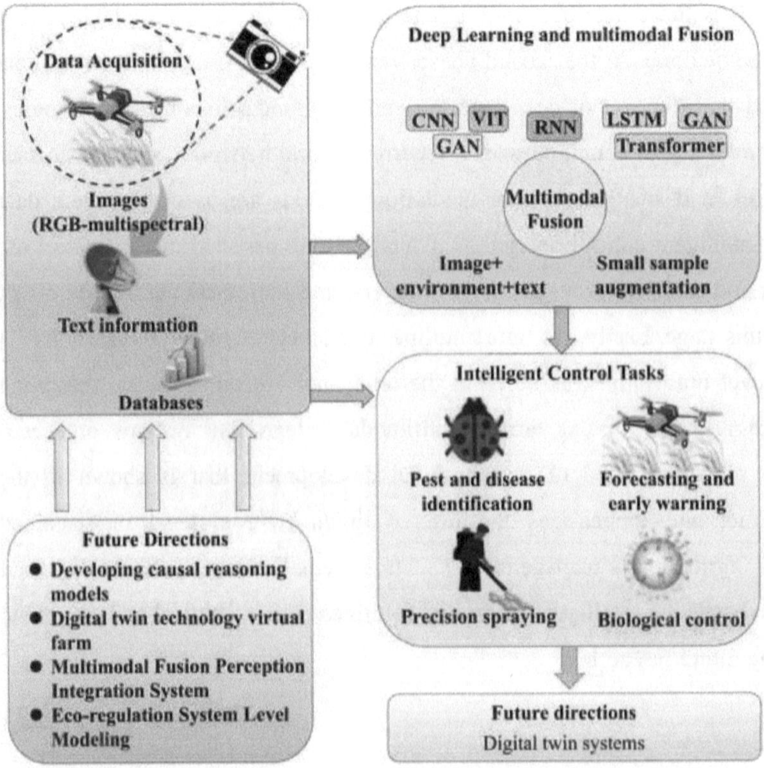

Figure 7. Application workflow of deep learning for intelligent pest and disease monitoring and control.

On this basis, deep learning is also being incorporated into smart equipment, thus, advancing the development of technologies of agricultural pest and disease management [87]. Using artificial intelligence, sensor networks, and data analytics, scientists come up with smart surveillance systems that have all-weather monitoring features. Large scale installation of high-precision sensors and cameras on farmlands will help such systems to record real-time information on the status of crop growth, climatic changes, and the initial signs of the presence of pests and diseases and send real-time notifications that can be used to prevent or control them [29]. Deep learning technologies contribute to pest and disease detection and control entering a new age of increased intelligence, which is supported by artificial intelligence technologies. The automated monitoring, accurate identification and smart prediction enabled by AI makes the management of agricultural pests and diseases more efficient, environmental-friendly, and sustainable. Artificial Intelligence in the Pest and Disease Identification and Detection in Agriculture. Deep Learning Architectures of the Mainstream. Such sophisticated deep-learned models can be characterized by a high potential of application and excellent outcomes in farm-related practices such as identifying agricultural pests and diseases. Figure 8 depicts a schematic comparison of mainstream deep-learning structures used in agricultural pest and disease detection. Convolutional neural networks are the most popular network to apply to image analysis in agriculture pests and diseases. Each CNN performs convolutional and pooling layers sequentially, which automatically identify hierarchical spatial information within crop images including edges, lesion patterns and texture details. These characteristics are subsequently subjected to fully connected layers in which the classification or regression takes place, thus being able to recognize diseased leaves and fruits precisely [88, 89]. Recurrent neural networks, on the other hand, are more applicable to sequential agricultural data, e.g., disease progression curves or climatic time series. Their recurrent nature allows transfer of hidden states across time steps thus architecture a temporal dependency. More sophisticated versions like LSTM further increase the ability to capture long term effects in dynamic agricultural setting. Transformers are a recent invention that uses self-attention as a model of global dependencies, including all input factors in parallel.

This architecture does not just speed up the calculation, but it also enhances contextual representation, and Transformers are very useful to combine multi-source agricultural data, such as spectral, phenotypic, and environmental data [90]. The use of these deep-learning technologies has transformed the agricultural pest and disease detection system to more of an integrated intelligent analysis as opposed to the previous analysis performed on the static image, which includes the environmental variables, crop growth status, and previous data and therefore significantly enhances the accuracy and flexibility of the monitoring. Relative to conventional pest and disease recognition methods which depended upon handcrafted characteristics or manual scouting, mainstream deep-learning structures have delivered significant advances in the precision of classification, scale, and automation. As an example, CNN-based image classifiers are always 10-20 percent more accurate than the classical texture or color-based algorithms on the benchmark data, and RNNs and Transformers allow modeling the time-dependent and multi-modal data that cannot be represented by the traditional statistical methods [91]. Nevertheless, these improvements also point to some knowledge gaps. The models of deep-learning usually need big, well-labeled datasets, and their operation can reduce significantly when applied to the heterogeneous field conditions, including unstable light, occlusion, or background interference. Furthermore, they are limited in terms of computational costs restricted to use in low-power mobile devices that are widely used in agriculture.

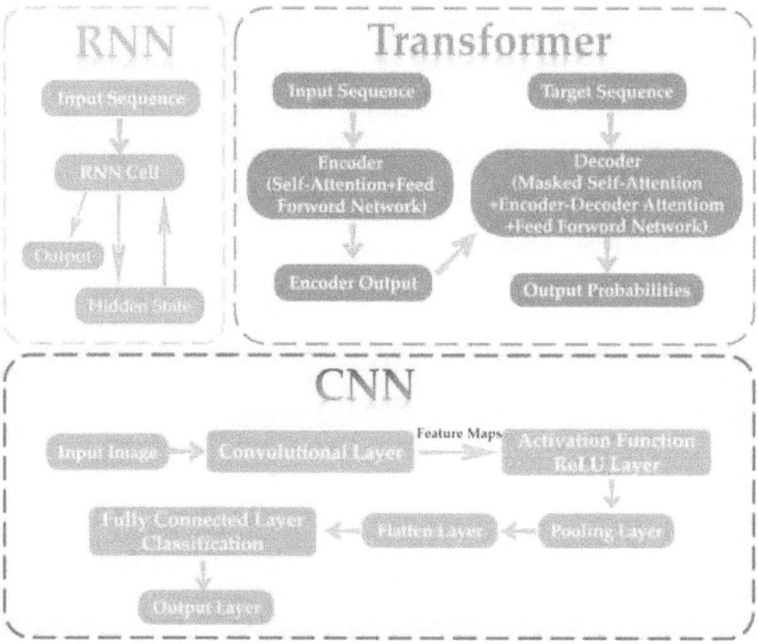

Figure 8. Schematic comparison of mainstream deep learning architectures (CNN, RNN, and Transformer) applied in agricultural pest and disease detection.

The development of SE -ResNet50 model, as depicted in Figure 9A, was implemented by using data augmentation and weight transfer, followed by optimizing the existing ResNet50 model, and used to detect powdery mildew on tomatoes. At the same time, the Transformer model exhibits outstanding global feature modeling functions in image recognition. Figure 9B illustrates that the model takes image data through a self-attention mechanism, which identifies salient features and produces attention maps to detect pests and diseases [92]. Besides the traditional Convolutional Neural Network and Transformer models, there has also been a significant potential in hyperspectral imaging to identify early signs of physiological disorders in fruits. Indicatively, Cong et al. [93] examined the chilling injury in kiwifruit, a disease that cannot easily be observed within the naked eye before a serious damage occurs. In order to resolve the issue of cross-variety variability of spectral properties, they proposed a general hybrid deep network (CDGSA-Net) optimized with a pelican optimization algorithm. The model attained accuracy, precision, recall, specificity, and F1 -score of over 99 percent, and thus, markedly performed better than traditional machine-learning and deep-learning baselines. In addition to demonstrating that hyperspectral imaging can be technically integrated with advanced deep learning, this paper also shows that local orchard crops like kiwifruit are practically relevant to localized detection since precision management is essential in small-

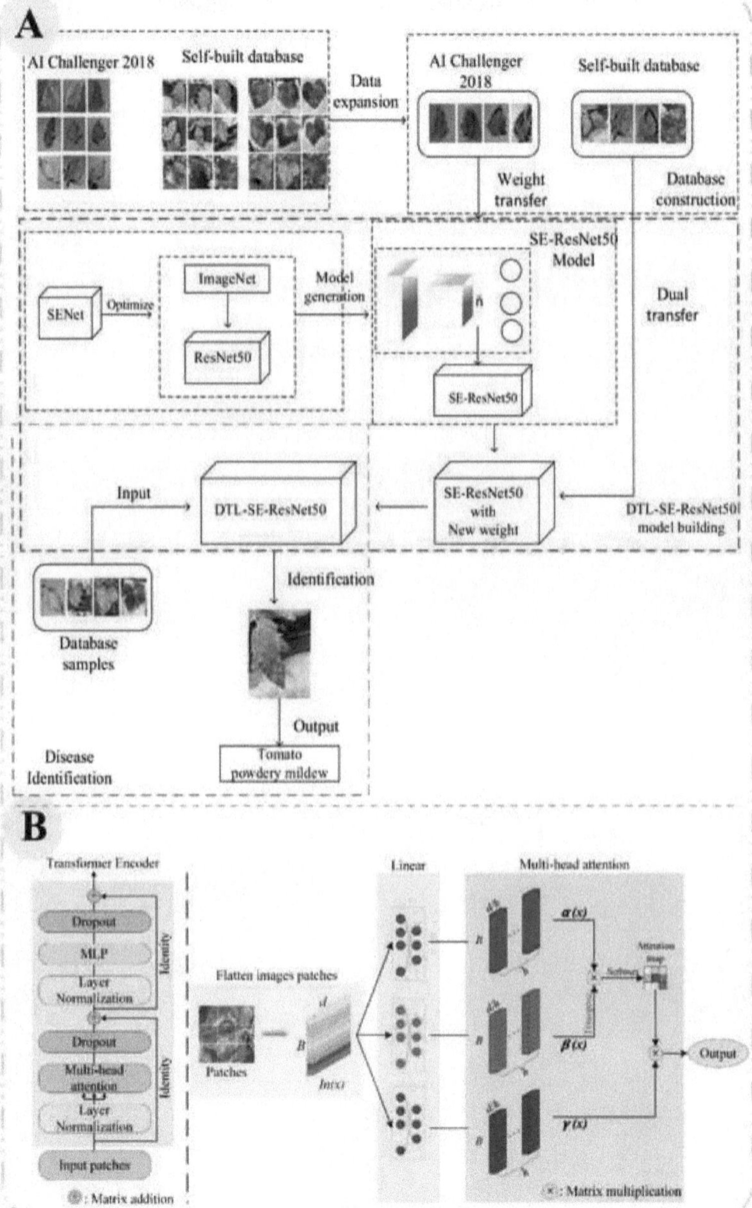
A
AI Challenger 2018
Self-built database
Data expansion
AI Challenger 2018
Self-built database
Weight transfer
Database construction
SE-ResNet50 Model
SENet
Optimize
ImageNet
ResNet50
Model generation
SE-ResNet50
Dual transfer
Input
DTL-SE-ResNet50
SE-ResNet50 with New weight
DTL-SE-ResNet50 model building
Identification
Database samples
Output
Tomato powdery mildew
Disease Identification
B
Transformer Encoder
Dropout
MLP
Layer Normalization
Identity
Dropout
Multi-head attention
Layer Normalization
Identity
Input patches
: Matrix addition
Flatten images patches
Patches
Linear
Multi-head attention
α(x)
β(x)
γ(x)
Softmax
Output
: Matrix multiplication

scales.

Figure 9. (**A**) A transfer learning workflow for image recognition and detection based on an improved ResNet50 combined with SENet for optimized feature extraction in deep learning; (**B**) A deep learning workflow for image recognition based on a Transformer architecture that extracts deep image features through a multi-head attention mechanism.

Significant advancements have been achieved in the application of image recognition and detection technologies for agricultural pest and disease management. Continuous innovations in model design and the fusion of diverse imaging modalities have accelerated the shift from manual diagnosis to automated and intelligent pest detection [**94**].

Figure 9. (A) A transfer-learning workflow of image recognition and detection based on an enhanced ResNet50 with SENet to extract deep image features through a multi-head attention mechanism; (B) An improved deep-learning workflow of image recognition based on a Transformer architecture to extract deep image features through a multi-head attention mechanism. Major advances have been made in the use of image-recognition and detection systems in the control of agricultural diseases and pests. The ongoing advances in the model design and integration of various imaging modalities have hastened the shift towards automated and smart pest detection as opposed to a manual diagnosis. UAV Image Analysis Technology Remote Sensing. The UAVs and satellite remote-sensing technologies are the sky-eye system of intelligent identification of agricultural pests and diseases and create a multi-scale monitoring network due to the collaboration of high and low altitudes. Using their low-altitude and high-resolution observation capabilities, UAVs are not limited by the capacities of satellites to atmospheric disturbance and their frequency of change, and satellites provide macro data at a broad scale and periodically. The combination of both can make the coverage of both individual plants to the regional scopes, which contributes greatly to the spatiotemporal effectiveness of early warning, dynamic disaster evaluation, and the sensitive control of pests and diseases. Satellite remote-sensing technology at the regional level is vital in monitoring the pests and diseases. However, the satellite imaging is still inferior to UAV imaging in terms of time and space with respect, especially in the areas of crop-field monitoring on a small scale where it might not provide enough precise results. As a result, UAV and satellite remote sensing should be integrated to provide a way of compensating their respective drawbacks even though data synchronization and fusion problems exist. In spite of the mentioned difficulties, technological advancements in remote-sensing surveillance have significantly contributed to the emergence of precision agriculture. There was a new method of biomass-prediction at the field scale using a multimodal remote-sensing system combined with terrain factors into a UAV-based device proposed by Zhang et al., which demonstrated better accuracy with biomass-prediction. Practically, the most essential element to attain precision agriculture is the effective combination of different remote-sensing data and the reduction of the effect of environmental factors. UAV-satellite joint remote sensing architecture(s) based on deep-learning models (CNN, RNN/LSTM, Transformer) to observe pests and diseases in the agricultural sector (Figure 10). The framework is made up of four components which include data acquisition (satellite reconnaissance and UAV surveys), data processing and fusion (pre-processing and multi-scale data integration), AI-powered analysis (CNN, RNN/LSTM, Transformer) and application in pest and disease detection, mapping, and early warning.

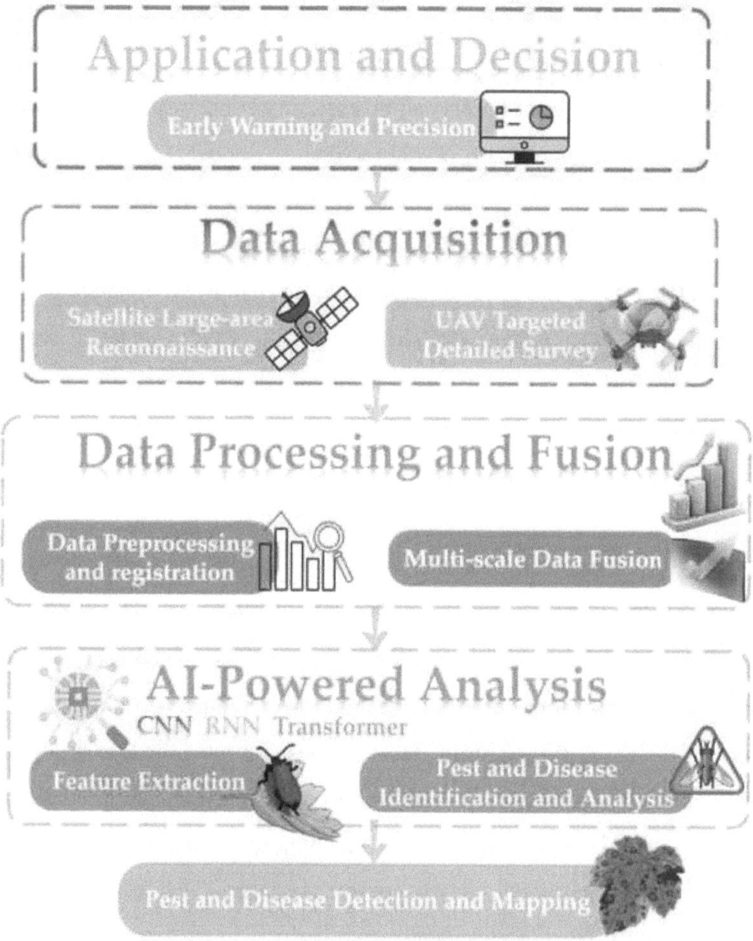

Figure 10. Framework of UAV–satellite collaborative remote sensing integrated with deep learning models (CNN, RNN/LSTM, Transformer) for agricultural pest and disease monitoring.

Conclusions

The past years have seen the revolution of deep learning technology in the intelligent detection and control of pests and diseases in agriculture. To begin with, in terms of pest and disease detection, visual recognition models that are based on deep learning have significantly improved both efficiency and accuracy of the identification. They can be used together with unmanned aerial vehicle remote sensing and satellite imagery to monitor large scale farmlands dynamically in real time. Second, the use of lightweight models in mobile devices and IoT sensor networks has continued to spur the massive growth and use of early warning systems on pests and diseases. Multimodal data fusion is used to combine environmental parameters, meteorological information, and image features and thus significantly enhances diagnostic capacities and yield strength in complicated agricultural settings. Moreover, when it comes to the area of control, deep learning-based precision spraying systems (applying object detection and path-planning algorithms) can optimize the application procedure, which can not only enhance the rate of utilization of pesticides but also reduce pollution of the environment. Lastly, smart early warning technologies that integrate historical pest and disease data with information on the crop growth cycle can predict the probability of a pest and disease outbreak, thus helping the decision-makers in the development of more scientifically-based control measures. The technologies of disease-resistant gene screening and phenotypic analysis based on deep learning have also accelerated the breeding of resistant crop types, and the complex control over the biological control and the ecological regulation provided new opportunities to develop green agriculture.

References

1. Bauer, M.E. Remote Sensing of Environment: History, Philosophy, Approach and Contributions, 1969–2019. *Remote Sens. Environ.* **2020**, *237*, 111522.
2. Shi, J.; Wu, T.; Kai Qin, A.; Lei, Y.; Jeon, G. Self-Guided Autoencoders for Unsupervised Change Detection in Heterogeneous Remote Sensing Images. *IEEE Trans. Artif. Intell.* **2024**, *5*, 2458–2471.
3. Duan, P.; Wang, Y.; Yin, P. Remote Sensing Applications in Monitoring of Protected Areas: A Bibliometric Analysis. *Remote Sens.* **2024**, *12*, 772.
4. Janga, B.; Asamani, G.P.; Sun, Z.; Cristea, N. A Review of Practical AI for Remote Sensing in Earth Sciences. *Remote Sens.* **2023**, *15*, 4112.
5. Chamasemani, F.F.; Singh, Y.P. Multi-Class Support Vector Machine (SVM) Classifiers—An Application in Hypothyroid Detection and Classification. In Proceedings of the 2011 Sixth International Conference on Bio-Inspired Computing: Theories and Applications, Penang, Malaysia, 27–29 September 2011; pp. 351–356.
6. Islam, K.I.; Elias, E.; Carroll, K.C.; Brown, C. Exploring Random Forest Machine Learning and Remote Sensing Data for Streamflow Prediction: An Alternative Approach to a Process-Based Hydrologic Modeling in a Snowmelt-Driven Watershed. *Remote Sens.* **2023**, *15*, 39999.
7. Mitchell, A.L.; Rosenqvist, A.; Mora, B. Current Remote Sensing Approaches to Monitoring Forest Degradation in Support of Countries Measurement, Reporting and Verification (MRV) Systems for REDD+. *Carbon Balance Manag.* **2017**, *12*, 9.
8. Chen, Y.; Morton, D.C.; Randerson, J.T. Remote Sensing for Wildfire Monitoring: Insights into Burned Area, Emissions, and Fire Dynamics. *One Earth* **2025**, *7*, 1022–1028.

9. G. Ceballos, P. R. Ehrlich, R. Dirzo, Biological annihilation via the ongoing sixth mass extinction signaled by vertebrate population losses and declines. *Proc. Natl. Acad. Sci. U.S.A.* **114**, E6089–E6096 (2017).
10. M. Dornelas et al., BioTIME: A database of biodiversity time series for the Anthropocene. *Glob. Ecol. Biogeogr.* **27**, 760–786 (2018).
11. V. Elbrecht et al., Validation of COI metabarcoding primers for terrestrial arthropods. *PeerJ* **7**, e7745 (2019).
12. H. Yousif, J. Yuan, R. Kays, Z. He, Animal scanner: Software for classifying humans, animals, and empty frames in camera trap images. *Ecol. Evol.* **9**, 1578–1589 (2019).
13. J. Ärje et al., Human experts vs. machines in taxa recognition. *Signal Process. Image Commun.* **87**, 115917 (2020).
14. S. Christin, É. Hervet, N. Lecomte, Applications for deep learning in ecology. *Methods Ecol. Evol.* **10**, 1632–1644 (2019).
15. I. Ruczyński, Z. Hałat, M. Zegarek, T. Borowik, D. K. N. Dechmann, Camera transects as a method to monitor high temporal and spatial ephemerality of flying nocturnal insects. *Methods Ecol. Evol.* **11**, 294–302 (2020).
16. S. E. Barlow, M. A. O'Neill, Technological advances in field studies of pollinator ecology and the future of e-ecology. *Curr. Opin. Insect Sci.* **38**, 15–25 (2020).
17. I. Potamitis, P. Eliopoulos, I. Rigakis, Automated remote insect surveillance at a global scale and the internet of things. *Robotics* **6**, 19 (2017).
18. I. Potamitis, I. Rigakis, N. Vidakis, M. Petousis, M. Weber, Affordable bimodal optical sensors to spread the use of automated insect monitoring. *J. Sens.* **2018**, 3949415 (2018).

19. D. J. A. Rustia, J.-J. Chao, J.-Y. Chung, T.-T. Lin, "An online unsupervised deep learning approach for an automated pest insect monitoring system" in *2019 ASABE Annual International Meeting* (ASABE, St. Joseph, MI, 2019), pp. 1–13.
20. T. M. Poland, D. Rassati, Improved biosecurity surveillance of non-native forest insects: A review of current methods. *J. Pest. Sci.* **92**, 37–49 (2019).
21. D. A. A. Santos, L. E. Teixeira, A. M. Alberti, V. Furtado, J. J. P. C. Rodrigues, "Sensitivity and noise evaluation of an optoelectronic sensor for mosquitoes monitoring" in *2018 3rd International Conference on Smart and Sustainable Technologies* (SpliTech, Split, Croatia, 2018), pp. 1–5.
22. J. Park, D. I. Kim, B. Choi, W. Kang, H. W. Kwon, Classification and morphological analysis of vector mosquitoes using deep convolutional neural networks. *Sci. Rep.* **10**, 1012 (2020).
23. R. Kalamatianos, I. Karydis, D. Doukakis, M. Avlonitis, DIRT: The dacus image recognition toolkit. *J. Imaging* **4**, 129 (2018).
24. W. Ding, G. Taylor, Automatic moth detection from trap images for pest management. *Comput. Electron. Agric.* **123**, 17–28 (2016).
25. T. T. Høye, H. M. R. Mann, K. Bjerge, "Camera-based monitoring of insects on green roofs" (Rep. no. 371, Aarhus University DCE–National Centre for Environment and Energy, Aarhus, Denmark, 2020).
26. K. Bjerge, M. V. Sepstrup, J. B. Nielsen, F. Helsing, T. T. Høye, An automated light trap to monitor moths (Lepidoptera) using computer vision-based tracking and deep learning. bioRxiv:10.1101/2020.03.18.996447 (31 October 2020).
27. J. W. Chapman, V. A. Drake, D. R. Reynolds, Recent insights from radar studies of insect flight. *Annu. Rev. Entomol.* **56**, 337–356 (2011).

28. O. Hueppop et al., Perspectives and challenges for the use of radar in biological conservation. *Ecography* **42**, 912–930 (2019).
29. K. R. Wotton et al., Mass seasonal migrations of hoverflies provide extensive pollination and crop protection services. *Curr. Biol.* **29**, 2167–2173.e5 (2019).
30. P. M. Stepanian et al., Declines in an abundant aquatic insect, the burrowing mayfly, across major North American waterways. *Proc. Natl. Acad. Sci. U.S.A.* **117**, 2987–2992 (2020).
31. A. Jeliazkov et al., Large-scale semi-automated acoustic monitoring allows to detect temporal decline of bush-crickets. *Glob. Ecol. Conserv.* **6**, 208–218 (2016).
32. I. Kiskin et al., Bioacoustic detection with wavelet-conditioned convolutional neural networks. *Neural Comput. Appl.* **32**, 915–927 (2020).
33. E. D. Chesmore, E. Ohya, Automated identification of field-recorded songs of four British grasshoppers using bioacoustic signal recognition. *Bull. Entomol. Res.* **94**, 319–330 (2004).
34. S. Kawakita, K. Ichikawa, Automated classification of bees and hornet using acoustic analysis of their flight sounds. *Apidologie (Celle)* **50**, 71–79 (2019).
35. V. Blagoderov, I. J. Kitching, L. Livermore, T. J. Simonsen, V. S. Smith, No specimen left behind: Industrial scale digitization of natural history collections. *ZooKeys*, 133–146 (2012).
36. B. Ströbel, S. Schmelzle, N. Blüthgen, M. Heethoff, An automated device for the digitization and 3D modelling of insects, combining extended-depth-of-field and all-side multi-view imaging. *ZooKeys* **759**, 1–27 (2018).
37. A. E. Z. Short, T. Dikow, C. S. Moreau, Entomological collections in the age of big data. *Annu. Rev. Entomol.* **63**, 513–530 (2018).

38. E. K. Meineke, T. J. Davies, Museum specimens provide novel insights into changing plant-herbivore interactions. *Philos. Trans. R. Soc. Lond. B Biol. Sci.* **374**, 20170393 (2018).
39. E. K. Meineke, C. Tomasi, S. Yuan, K. M. Pryer, Applying machine learning to investigate long-term insect-plant interactions preserved on digitized herbarium specimens. *Appl. Plant Sci.* **8**, e11369 (2020).
40. J. Ärje et al., Automatic image-based identication and biomass estimation of invertebrates. *Methods Ecol. Evol.* **11**, 922–931 (2020).
41. W. Luo, X. Zhao, T.-K. Kim, Multiple object tracking: A review. arXiv:1409.7618 (26 September 2014).
42. B. Yang, R. Nevatia, "Multi-target tracking by online learning of non-linear motion patterns and robust appearance models" in *2012 IEEE Conference on Computer Vision and Pattern Recognition (CVPR)* (IEEE, 2012), pp. 1918–1925.
43. D. T. Tran, S. Kiranyaz, M. Gabbouj, A. Iosifidis, Heterogeneous multilayer generalized operational perceptron. *IEEE Trans. Neural Netw. Learn. Syst.* **31**, 710–724 (2020).
44. D. T. Tran, A. Iosifidis, M. Gabbouj, Improving efficiency in convolutional neural networks with multilinear filters. *Neural Netw.* **105**, 328–339 (2018).
45. A. Valiente-Banuet et al., Beyond species loss: The extinction of ecological interactions in a changing world. *Funct. Ecol.* **29**, 299–307 (2015).
46. S. Hamel et al., Towards good practice guidance in using camera-traps in ecology: Influence of sampling design on validity of ecological inferences. *Methods Ecol. Evol.* **4**, 105–113 (2013).
47. L. Estes et al., The spatial and temporal domains of modern ecology. *Nat. Ecol. Evol.* **2**, 819–826 (2018).

48. J. Raitoharju, K. Meissner, "On confidences and their use in (semi-)automatic multi-image taxa identification" in *2019 IEEE Symposium Series on Computational Intelligence (SSCI)* (IEEE, 2019), pp. 1338–1343.
49. M. Valan, K. Makonyi, A. Maki, D. Vondráček, F. Ronquist, Automated taxonomic identification of insects with expert-level accuracy using effective feature transfer from convolutional networks. *Syst. Biol.* **68**, 876–895 (2019).
50. D. Milošević et al., Application of deep learning in aquatic bioassessment: Towards automated identification of non-biting midges. *Sci. Total Environ.* **711**, 135160 (2020).
51. G. V. Horn et al., The iNaturalist challenge 2017 dataset. arXiv:1707.06642 (20 July 2017).
52. O. L. P. Hansen et al., Species-level image classification with convolutional neural network enables insect identification from habitus images. *Ecol. Evol.* **10**, 737–747 (2020).
53. Y. LeCun, Y. Bengio, G. Hinton, Deep learning. *Nature* **521**, 436–444 (2015).
54. A. Joly et al., "Overview of LifeCLEF 2018: A large-scale evaluation of species identification and recommendation algorithms in the Era of AI" in *Experimental IR Meets Multilinguality, Multimodality, and Interaction*, P. Bellot et al., Eds. (Springer International Publishing, Cham, Switzerland, 2018), pp. 247–266.
55. C. Huang, Y. N. Li, C. C. Loy, X. O. Tang, "Learning deep representation for imbalanced classification" in *2016 IEEE Conference on Computer Vision and Pattern Recognition (CVPR)* (IEEE, 2016), pp. 5375–5384.
56. X. Y. Jing et al., Multiset feature learning for highly imbalanced data classification. *IEEE Trans. Pattern Anal. Mach. Intell.*, (2019).

57. J. Raitoharju et al., Benchmark database for fine-grained image classification of benthic macroinvertebrates. *Image Vis. Comput.* **78**, 73–83 (2018).
58. M. Turkoz, S. Kim, Y. Son, M. K. Jeong, E. A. Elsayed, Generalized support vector data description for anomaly detection. *Pattern Recognit.* **100**, 107119 (2020).
59. P. Perera, V. M. Patel, "Deep transfer learning for multiple class novelty detection" in *2019 IEEE Conference on Computer Vision and Pattern Recognition (CVPR)* (IEEE, 2019), pp. 11536–11544.
60. C. Geng, S. J. Huang, S. Chen, Recent advances in open set recognition: A survey. *IEEE Trans. Pattern Anal. Mach. Intell.*, (2020).
61. P. F. Thomsen, E. E. Sigsgaard, Environmental DNA metabarcoding of wild flowers reveals diverse communities of terrestrial arthropods. *Ecol. Evol.* **9**, 1665–1679 (2019).
62. A. Kendall, Y. Gal, R. Cipolla, "Multi-task learning using uncertainty to weigh losses for scene geometry and semantics" in *2018 IEEE Conference on Computer Vision and Pattern Recognition (CVPR)* (IEEE, 2018), pp. 7482–7491.
63. L. A. Gatys, A. S. Ecker, M. Bethge, "Image style transfer using convolutional neural networks" in *2016 IEEE Conference on Computer Vision and Pattern Recognition (CVPR)* (IEEE, 2016), pp. 2414–2423.
64. E. Tzeng, J. Hoffman, K. Saenko, T. Darrell, "Adversarial discriminative domain adaptation" in *2017 IEEE Conference on Computer Vision and Pattern Recognition (CVPR)* (IEEE, 2017), pp. 2962–2971.
65. P. Glover-Kapfer, C. A. Soto-Navarro, O. R. Wearn, Camera-trapping version 3.0: Current constraints and future priorities for development. *Remote Sens. Ecol. Conserv.* **5**, 209–223 (2019).

66. M. Willi et al., Identifying animal species in camera trap images using deep learning and citizen science. *Methods Ecol. Evol.* **10**, 80–91 (2019).

67. Y. Zhong, J. Gao, Q. Lei, Y. Zhou, A vision-based counting and recognition system for flying insects in intelligent agriculture. *Sensors (Basel)* **18**, 1489 (2018).

68. H. Inoue, Data augmentation by pairing samples for images classification. arXiv:1801.02929 (9 January 2018).

69. Khajuria, R., Sunanda, D., Singh, S. A survey of existing leaf disease techniques using artificial neural network. *Int. J. Comput. Trends Technol.*, **59**(1), 52-62. https://doi.org/10.14445/22312803/IJCTT-V59P109 (2018).

70. Atalla, S. et al. IoT-enabled precision agriculture: Developing an ecosystem for optimized crop management. *Information* **14**(4), 205. https://doi.org/10.3390/info14040205 (2023).

71. Li, Y. & Chao, X. Ann-based continual classification in agriculture. *Agriculture* **10**(5), 178. https://doi.org/10.3390/agriculture10050178 (2020).

72. Hahn, F., Valle, S., Rendón, R., Oyorzabal, O. & Astudillo, A. Mango fruit fly trap detection using different wireless communications. *Agronomy* **13**(7), 1736. https://doi.org/10.3390/agronomy13071736 (2023).

73. Tian, E., Li, Z., Huang, W. & Ma, H. Distributed and parallel simulation methods for pest control and crop monitoring with IoT assistance. *Acta Agric. Scand. Sect. B Soil Plant Sci.* **71**(9), 884–898. https://doi.org/10.1080/09064710.2021.1955959 (2021).

74. Dang, L. M., Piran, M. J., Han, D., Min, K. & Moon, H. A survey on internet of things and cloud computing for healthcare. *Electronics* **8**(7), 768. https://doi.org/10.3390/electronics8070768 (2019).
75. Lengai, G. M. W., Muthomi, J. W. & Mbega, E. R. Phytochemical activity and role of botanical pesticides in pest management for sustainable agricultural crop production. *Sci. Afr.* **7**, e00239. https://doi.org/10.1016/j.sciaf.2019.e00239 (2020).
76. Wang, F., Wang, R., Xie, C., Yang, P. & Liu, L. Fusing multi-scale context-aware information representation for automatic in-field pest detection and recognition. *Comput. Electron. Agric.* **169**, 105222. https://doi.org/10.1016/j.compag.2020.105222 (2020).
77. Wang, Q. J. et al. Pest24: A large-scale very small object data set of agricultural pests for multi-target detection. *Comput. Electron. Agric.* **175**, 105585. https://doi.org/10.1016/j.compag.2020.105585 (2020).
78. Li, Y., Wang, H., Dang, L. M., Sadeghi-Niaraki, A. & Moon, H. Crop pest recognition in natural scenes using convolutional neural networks. *Comput. Electron. Agric.* **169**, 105174. https://doi.org/10.1016/j.compag.2019.105174 (2020).
79. Salamut, C., Kohnert, I., Landwehr, N., Pflanz, M., Schirrmann, M., Zare, M. Deep learning object detection for image analysis of cherry fruit fly (*Rhagoletis cerasi* L.) on yellow sticky traps. *Gesunde Pflanzen*, **75**, 37-48, https://doi.org/10.1007/s10343-022-00794-0 (2023).
80. Yang, Z. K., Li, W. Y., Li, M., Yang, X. T. Automatic greenhouse pest recognition based on multiple color space features. *Int. J .Agric. Biol. Eng.*, **14**(2), 188–195, https://doi.org/10.25165/j.ijabe.20211402.5098 (2021).

81. Yihao, W., Chunjiang, Z., Daming, D. & Kun, W. Real-time monitoring of insects based on laser remote sensing. *Ecol. Indicat.* **151**, 110302. https://doi.org/10.1016/j.ecolind.2023.110302 (2023).
82. Kathole, A. B., Vhatkar, K. N., Kumbhare, S., Katti, J. & Kimbahune, V. V. IoT-based smart agriculture for onion plant disease management: A comprehensive approach. *Int. J. Intell. Syst. Appl. Eng.* **12**(3), 472–476 (2024).
83. Kathole, A. B., Katti, J., Ubale, S. A., Vhatkar, K. N., Jadhav, D., & Kimbahune, V. V. IoT approach for ensuring safety in smart cities for energy saving. *J. Electr. Syst.*, **20**(2), 418–423. https://doi.org/10.52783/jes.1337
84. Too, E.C.; Yujian, L.; Njuki, S.; Yingchun, L. A comparative study of fine-tuning deep learning models for plant disease identification. *Comput. Electron. Agric.* **2019**, *161*, 272–279.
85. Ripper, W.E. Biological control as a supplement to chemical control of insect pests. *Nature* **1944**, *153*, 448–452.
86. Stenberg, J.A.; Sundh, I.; Becher, P.G.; Björkman, C.; Dubey, M.; Egan, P.A.; Friberg, H.; Gil, J.F.; Jensen, D.F.; Jonsson, M. When is it biological control? A framework of definitions, mechanisms, and classifications. *J. Pest. Sci.* **2021**, *94*, 665–676
87. Wang, T.; Zhao, L.; Li, B.; Liu, X.; Xu, W.; Li, J. Recognition and counting of typical apple pests based on deep learning. *Ecol. Inform.* **2022**, *68*, 101556.
88. Ahanger, T.A.; Bhatia, M.; Alabduljabbar, A.; Albanyan, A. Blockchain-based intelligent equipment assessment in manufacturing industry. *J. Intell. Manuf.* **2025**, 1–19.
89. Li, H.; Guo, Y.; Zhao, H.; Wang, Y.; Chow, D. Towards automated greenhouse: A state of the art review on greenhouse monitoring methods and

technologies based on internet of things. *Comput. Electron. Agric.* **2021**, *191*, 106558.

90. Qiu, D.; Guo, T.; Yu, S.; Liu, W.; Li, L.; Sun, Z.; Peng, H.; Hu, D. Classification of apple color and deformity using machine vision combined with cnn. *Agriculture* **2024**, *14*, 978.
91. Pei, E.; Zhao, Y.; Oveneke, M.C.; Jiang, D.; Sahli, H. A Bayesian filtering framework for continuous affect recognition from facial images. *IEEE Trans. Multimed.* **2022**, *25*, 3709–3722.
92. Xu, C.; Sui, X.; Liu, J.; Fei, Y.; Wang, L.; Chen, Q. Transformer in optronic neural networks for image classification. *Opt. Laser Technol.* **2023**, *165*, 109627.
93. Cong, S.; Sun, J.; Shi, L.; Dai, C.; Wu, X.; Zhang, B.; Yao, K. Hyperspectral imaging combined with a universal hybrid deep network for identifying early chilling injury in kiwifruit across varieties. *Postharvest Biol. Technol.* **2025**, *230*, 113752.
94. Zhao, S.; Peng, Y.; Liu, J.; Wu, S. Tomato leaf disease diagnosis based on improved convolution neural network by attention module. *Agriculture* **2021**, *11*, 651.
95. Feng, G.; Wang, C.; Wang, A.; Gao, Y.; Zhou, Y.; Huang, S.; Luo, B. Segmentation of Wheat Lodging Areas from UAV Imagery Using an Ultra-Lightweight Network. *Agriculture* **2024**, *14*, 244.

About Authors

Prof. Dr. Mohamed Abdel-Raheem Ali Abdel-Raheem

Professor of Entomology (Biological Control)

Pests & Plant Protection Department

Agricultural and Biological Research Institute

National Research Centre.

33rd ElBohouth St., Dokki, Cairo, Egypt.

Email: abdelraheem_nrc@hotmail.com,

abdelraheem_nrc@yahoo.com

Mobile: (+2) 01155527583 - (+2) 01009580797

Prof. Dr. Mohamed Abdel-Raheem Ali Abdel-Raheem, Prof. of Entomology (Biological Control), Pests & Plant Protection Department, Agricultural and Biological Research Institute, National Research Centre, Cairo, Egypt. Published publications (323) papers (88), Books and Book chapters (235), Edited books (8), (Scopus) h-index (12), Citations (341), (Google Scholar) h-index (21) , Citations (1212), (Web of Science) Publications (29), h-index (11), Citations (318), Research Gate, h-index (15), Citations (734), Reviewers in International Journals (160), Reviewed articles in Biological Control (1012), Editor in Chief in Journal (7), Assistant Editor (1), Associated Editor in Chief in Journal (7), Editorial board in Journals (43), Advisory board (1), conference Attended (27), workshop Attended (313), Symposium Attended (179), Forum Attended (4), Prize (4), Attended Others (30), Projects As PI and member (20), Training courses For Agric. Engineering (61), Training courses For Students of University (12), attending Trainees courses (19), TV, Radio & Newspaper (39), Supervisor on Ms.C. thesis (3), Supervisor on Ph.D. thesis (1), Committee Ph.D. Thesis in Egypt (2), and member in scientific Foundation (14), Giving International lectures (4), Promotion Committee Prof. international (2), Committee Ph.D. thesis international (1).

National Research Centre. 33rd ElBohouth St., Dokki, Cairo, Egypt.

Email Address: abdelraheem_nrc@hotmail.com, ma.abdel-raheem@nrc.sci.eg, abdelraheem_nrc@yahoo.com, profmohamedabdelraheem@gmail.com, abdelraheem196942@gmail.com, mabdelraheem69@gmail.com, mabdelraheem69@yahoo.com.

Mobile Phone: (+2) 01155527583 – (+2) 01009580797

Scopus Author ID : 55220661700

https://www.scopus.com/authid/detail.uri?authorId=55220661700

http://scholar.google.com/citations?hl=en&user=gBtssEgAAAAJ

https://orcid.org/my-orcid?orcid=0000-0001-9240-064X

https://www.webofscience.com/wos/author/searchAbdel-Raheem, Mohamed Abdel-Raheem Ali - Web of Science Core Collection

https://www.researchgate.net/profile/Mohamed_Abdel-Raheem4/publications?sorting=recentlyAdded

http://livedna.org/20.16018

http://www.resarcherid.com/rid/R-6264-2017

https://www.webofscience.com/wos/author/record/R-6264-2017

https://www.facebook.com/share/p/1Piu6xyFmN/

- TV, Radio & Newspaper links

1. https://www.youtube.com/attribution_link?a=NsH2xGv9mY8&u=%2Fwatch%3Fv%3Dsb46rBG8-Mc%26feature%3Dshare
2. https://youtu.be/Ub8djW0tggo
3. https://m.facebook.com/story.php?story_fbid=492907069288338&id=100076011247165
4. https://youtu.be/dOujUAZmTdw
5. https://m.facebook.com/story.php?story_fbid=492907069288338&id=100076011247165
6. https://fb.watch/gPGrDuzZFX/
7. https://www.youtube.com/watch?si=4UfyQaeHVLPiVR5J&v=nXV4NT-no7o&feature=youtu.be
8. https://www.youtube.com/watch?v=sRH44W6TbcQ
9. https://youtu.be/uq9Qj3AEBHk?si=ymaX06huG2AXJESu

10. https://www.facebook.com/share/v/Qg5s6HW3hD8PLBzp/?mibextid=xfxF2i
11. https://el-yom.com/news/archives/1092767
12. https://el-yomnews.com/?p=1156515
13. https://el-yomnews.com/%D8%A7%D9%84%D8%AF%D9%83%D8%AA%D9%88%D8%B1-%D8%B9%D8%A8%D8%AF%D8%A7%D9%84%D8%B1%D8%AD%D9%8A%D9%85-%D9%85%D8%AD%D9%85%D8%AF-%D8%B9%D9%84%D9%8A-%D9%8A%D8%AA%D8%AD%D8%AF%D8%AB-%D9%84-%D9%84%D9%84/
14. https://el-yomnews.com/?p=1179515
15. https://www.facebook.com/share/v/4MM85wKRPxsYLqTr/
16. http://https//el-yomnews.com/%D8%A7%D9%84%D8%AD%D8%B4%D8%B1%D8%A7%D8%AA-%D9%81%D9%8A-%D8%AE%D8%AF%D9%85%D8%A9-%D8%A7%D9%84%D8%B9%D8%AF%D8%A7%D9%84%D8%A9-%D9%85%D8%AD%D8%A7%D8%B6%D8%B1%D8%A9-%D8%AA%D9%83%D8%B4%D9%81-%D8%AF%D9%88/
17. https://youtube.com/watch?v=F21ziQxdeco&si=LvSUixmnSg4JM1zK
18. https://youtu.be/F21ziQxdeco?si=jyAwbU8dVwkYzRH4
19- https://el-yomnews.com/المكافحة-الحيوية-بالفطريات-حلول-مبتك/
20- https://el-yomnews.com/archives/1296404
21- https://www.facebook.com/share/p/1AC25ue5XG/
22- https://youtu.be/7VItE5kON2g?si=9rMWMUSGk8Csgjfn
23- https://www.youtube.com/watch?v=G9nXA7Q9C60

24- https://www.youtube.com/watch?v=7VItE5kON2g

25- https://el-yomnews.com/archives/1347977

26- https://youtu.be/x9N58DakONg?si=Uh0HbpQsiDRpIGMl

27- https://youtu.be/PfuQy-Zkq3s?si=GtRRxy9XtTW-YgF3

28- https://youtu.be/Vo1NqVmPxMU?si=vJxt4YyhzqJL9hR4

29- https://youtu.be/pqo5F-cbGSc?si=J_B9Dv05bfX5P-7O

TV and Radio

1- Filming in the program (Meeting the Scientists) - Al-Manara Channel, entitled Biological Control of Insect Pests, on 7/2/2013.
2- Filming in the program (Friends of Al-Manara) - Al-Manara Channel, February 2014.
3- A recording of an interview with the program (From the Secrets of the Universe and Life) Radio Sawt Al-Arab, Cairo on 4/26/2015.
4- Filming in the program (Idea for a bobbin) - Nile Cultural Channel under the title "Using entomopathogenic fungi and plant extracts in combating insect pests that infect sugar crops in Egypt" on 11/1/2015.
5- Filming on (This Morning) program - Nile News Channel under the title: Submitting the specifications for the reclamation of one and a half million feddans as a first stage of the reclamation of 4 million feddans on 11/15/2015. https://www.youtube.com/attribution_link?a=NsH2xGv9mY8&u=%2Fwatch%3Fv%3Dsb46rBG8-Mc%26feature%3Dshare
6- Registration in the program (Our Life Between Our Hands) for North Upper Egypt Radio and Radio, presented by Dr. Essam Fathy, 3 episodes, on 11-29-

2021, and it will be broadcast on Saturday and Wednesday at 3:30 pm in January 2022

7- Registration in the program (Our Life Between Our Hands) for North Upper Egypt Radio and Radio, presented by Dr. Essam Fathy, 3 episodes, on 11-29-2021, and it will be broadcast on Saturday and Wednesday at 3:30 pm in January 2022

8- Registration in the program (Our Life Between Our Hands) for North Upper Egypt Radio and Radio, presented by Dr. Essam Fathy, 3 episodes, on 11-29-2021, and it will be broadcast on Saturday and Wednesday at 3:30 pm in January 2022

9- Registration in "Our Green Land" program for North Upper Egypt Radio and Radio, presented by Dr. Essam Fathy, 3 episodes, on 11-2021-2021, to be broadcast in January 2022

10- Registration in "Our Green Land" program for North Upper Egypt Radio and Radio, presented by Dr. Essam Fathy, 3 episodes, on 11-2021-2021, to be broadcast in January 2022

11- Registration in "Our Green Land" program for North Upper Egypt Radio and Radio, presented by Dr. Essam Fathy, 3 episodes, on 11-2021-2021, to be broadcast in January 2022

12- Filming in the program (Egypt is Beautiful) - the second Egyptian Nile Channel, under the title: Climate Changes on November 6, 2022. https://youtu.be/dOujUAZmTdw

13- Registration in "Our Green Land" program for North Upper Egypt Radio and Radio, presented by Dr. Essam Fathy, 3 episodes, on 11-2021-2021, to be broadcast in January 2022

14- Filming in the program (Egypt and Climate Summit) - Delta Satellite Channel under the title: Climate Cop 27 on 13/11/2022 https://youtu.be/Ub8djW0tggo

15- Filming in (Egypt) - Sohag, Coral Times Hotel, under the title: Climate Change, Cop 27, dated 11/15/2022, the journalist Shorouk Abdullah https://m.facebook.com/story.php?story_fbid=492907069288338&id=100076011247165

16- Filming in (Feed and Benefit) - Coral Hotel in Sohag under the title: Climate Cop 27 on 11/15/2022, journalist Ahmed Tarboush https://fb.watch/gPGrDuzZFX/

17- An intervention in the plant clinic program on insect-pathogenic fungi in Egypt's agricultural farms on Thursday 29/2/2024.
https://www.youtube.com/watch?si=4UfyQaeHVLPiVR5J&v=nXV4NT-no7o&feature=youtu.be

18- Filming in the plant clinic program on the Misr Agricultural Channel about insect-pathogenic fungi on Sunday 3-3-2024
https://www.youtube.com/watch?v=sRH44W6TbcQ

19- Filming in the plant clinic program at the Misr Agricultural Channel, early prediction of the red palm weevil and modern methods of combating it, on Thursday 25/4/2024.
https://youtu.be/uq9Qj3AEBHk?si=ymaX06huG2AXJESu

20- Filming in the Green Planet program on the Egyptian Channel 2 (Biological Control) on Sunday 7/6/2024
https://www.facebook.com/watch/?v=981454710052376

21- A press interview in Al-Youm newspaper with journalist Mr. Imad Naseer on 7-30-2024. https://el-yom.com/news/archives/1092767
https://el-yomnews.com/

22- https://el-yomnews.com/?p=1156515

23- https://el-yomnews.com/%D8%A7%D9%84%D8%AF%D9%83%D8%AA%D9%88%D8%B1-%D8%B9%D8%A8%D8%AF%D8%A7%D9%84%D8%B1%D8%AD%D9%8A%D9%85-%D9%85%D8%AD%D9%85%D8%AF-%D8%B9%D9%84%D9%8A-%D9%8A%D8%AA%D8%AD%D8%AF%D8%AB-%D9%84-%D9%84%D9%84/

24- https://el-yomnews.com/?p=1179515

25- https://www.facebook.com/share/v/4MM85wKRPxsYLqTr/

26- Insects in the Service of Justice: A Lecture Explaining the Role of Forensic Entomology in Investigating Crimes

https://el-yomnews.com/archives/1243894

27- Human health is at risk. An investigation reveals the secrets of pesticide-contaminated agriculture.

https://el-yomnews.com/archives/1296404

28- https://www.facebook.com/share/p/1AC25ue5XG/

The Green Planet Program (Beneficial Insects and Their Role in the Environment) on 5/23/2025 on the Egyptian Channel 2

29- https://youtu.be/7VItE5kON2g?si=9rMWMUSGk8Csgjfn

30- https://www.youtube.com/watch?v=G9nXA7Q9C60

31- https://www.youtube.com/watch?v=7VItE5kON2g

32- https://el-yomnews.com/archives/1347977

33- https://youtu.be/x9N58DakONg?si=Uh0HbpQsiDRpIGMl

34- https://youtu.be/PfuQy-Zkq3s?si=GtRRxy9XtTW-YgF3

35- https://youtu.be/Vo1NqVmPxMU?si=vJxt4YyhzqJL9hR4

36- https://youtu.be/pqo5F-cbGSc?si=J_B9Dv05bfX5P-7O

Assist. Prof. Eva A Ajaj

Internal and Preventive Medicine, Clinical Pathology, College of Veterinary Medicine, University of Mosul, Iraq

Lecture. Sajida Saeed Aziz

Department of Horticulture, College of Agricultural Engineering Sciences, University of Sulaimani, Iraq

Lecturer Fuad M. Ahmad

Department of Soil and water, College of Agriculture Engineering Sciences, Salahaddin University-Erbil, Iraq.

Dr. S.R. Rajkumar

The Head & Librarian at the Rev. Fr. S.J. Thomas EJ Library & Information Centre, Pilathara, affiliated with Kannur University, Kerala, India

Lecture. Viyan Ghazi Mohammed

Plant protection Department, College of Agricultural Engineering Science, Duhok University, Kurdistan, Iraq

FSC
www.fsc.org
MIX
Papier | Fördert gute Waldnutzung
FSC® C083411

Kolibri 360 GmbH
Ferdinand-Jühlke-Straße 7
99095 Erfurt, Deutschland
produktsicherheit@kolibri360.de

Druck:
CPI Druckdienstleistungen GmbH
im Auftrag der
Kolibri 360 GmbH
Ferdinand-Jühlke-Str. 7
99095 Erfurt